PENGUIN BOOKS

ANAXIMANDER

'Rovelli is a very good scientist and a very good writer . . . explaining some of the most conceptually difficult and densest areas of physics lightly and breezily . . . he tells the story of an ancient thinker who had a revolutionary idea about the Earth's place in the cosmos' Tom Whipple, *The Times*

'A celebration of the scientific spirit of inquiry and the remarkable achievements of one man more than 2,500 years ago' John Sellars, *TLS*

'A bold and persuasive case that this ancient Greek philosopher scientist was the founder of critical thinking' Adam Rutherford, *Start the Week*, BBC Radio 4

'This is seriously astounding. So lucid, so imaginative, so subtle, and so large in scope. It's like the best primer you can imagine for the non-scientist on why what you think you know about Ptolemy and Copernicus, or Popper and Kuhn, is not quite right' Sam Leith, *X*

'All Rovelli's wit, intellectual agility and most of his charm are in evidence in this thrilling early work' Simon Ings, *New Scientist*

'This book offers a timely rebuttal to those who would sacrifice the vital legacy of Western science' *Economist*

'What does set Rovelli's book apart is the irrepressible joy that bubbles through it. His summaries of Anaximander's astronomical thought are some of the most lucid and enthusiastic I've read' Claire Hall, *LRB*

'Rather thrilling . . . laudable and passionate' Steven Poole, *Spectator*

'An insightful survey of the scientific contributions of Greek philosopher Anaximander . . . Rovelli makes the most of the available evidence in building his case that the philosopher's emphasis on natural causes marked a sea change in human thought. This is a masterful overview of a pivotal figure in scientific history' *Publisher's Weekly*

ABOUT THE AUTHOR

Carlo Rovelli is a theoretical physicist who has made significant contributions to the physics of space and time. He has worked in Italy and the US and is currently directing the quantum gravity research group of the Centre de Physique Théorique in Marseille, France. His books *Seven Brief Lessons on Physics*, *Helgoland*, *Reality Is Not What It Seems* and *The Order of Time* are international bestsellers, which have been translated into forty-three languages.

CARLO ROVELLI

Anaximander

And the Nature of Science

Translated by Marion Lignana Rosenberg

PENGUIN BOOKS

PENGUIN BOOKS

UK | USA | Canada | Ireland | Australia
India | New Zealand | South Africa

Penguin Books is part of the Penguin Random House group of companies
whose addresses can be found at global.penguinrandomhouse.com.

Penguin Random House UK
One Embassy Gardens, 8 Viaduct Gardens, London SW11 7BW

penguin.co.uk

Penguin
Random House
UK

First published in French as *Anaximandre de Milet, ou la naissance
de la pensée scientifique* by Éditions Dunod 2009
English translation first published as *The First Scientist: Anaximander
and His Legacy* by Westholme Publishing 2016
Published in Penguin Books 2023
Published in paperback by Penguin Books 2025
001

Copyright © Carlo Rovelli, 2007

Printed and bound in Great Britain by Clays Ltd, Elcograf S.p.A.

The authorized representative in the EEA is Penguin Random House Ireland,
Morrison Chambers, 32 Nassau Street, Dublin D02 YH68

A CIP catalogue record for this book is available from the British Library

ISBN: 978-1-802-06304-2

Penguin Random House is committed to a sustainable future
for our business, our readers and our planet. This book is made from
Forest Stewardship Council® certified paper.

To Bonnie

CONTENTS

Rerum fores aperuisse, Anaximander Milesius traditur primus.

It is said that Anaximander of Miletus first opened the doors of nature.

—*Pliny, Natural History 2*

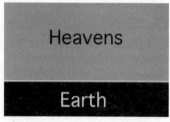 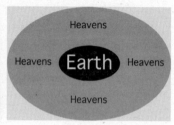

Figure 1a, left. Most early human civilizations viewed the world as the Heavens above and the Earth below. Figure 1b, right. The ancient Greeks saw the Earth as a stone floating in space.

INTRODUCTION

Human civilizations have always believed that the world consisted of the Heavens above and the Earth below (figure 1a). Beneath the Earth, to keep it from falling, there had to be more earth; or perhaps an immense turtle on the back of an elephant, as in some Asian myths; or gigantic columns like those supporting the Earth according to the Bible. This vision of the world was shared by the Egyptians, the Chinese, the Mayans, the peoples of ancient India and sub-Saharan Africa, the Hebrews, Native Americans, the ancient Babylonian empires, and all other cultures of which we have evidence.

All but one: the Greek world. Already in the classical era, the Greeks saw the Earth as a stone floating in space without falling (figure 1b). Beneath the Earth, there was neither more earth without limit, nor turtles, nor columns, but rather the same sky that we see over our heads. How did the Greeks manage to understand so early that the Earth is suspended in the void and that the Heavens continue under our feet? Who understood this, and how?

The man who made this enormous leap in understanding the world is the main character in this story:

Ἀναξίμανδρος, Anaximander, who lived twenty-six centuries ago in Miletus, a Greek city on the coast of what is now Turkey. This discovery alone would make Anaximander one of the intellectual giants of the ages. But Anaximander's legacy is still greater. He paved the way for physics, geography, meteorology, and biology. Even more important than these contributions, he set in motion the process of rethinking our worldview—a search for knowledge based on the rejection of any obvious-seeming "certainty," which is one of the main roots of scientific thinking.

The nature of scientific thinking is the second subject of this book. Science, I believe, is a passionate search for always newer ways to conceive the world. Its strength lies not in the certainties it reaches but in a radical awareness of the vastness of our ignorance. This awareness allows us to keep questioning our own knowledge, and, thus, to continue learning. Therefore the scientific quest for knowledge is not nourished by certainty, it is nourished by a radical lack of certainty. Its way is fluid, capable of continuous evolution, and has immense strength and a subtle magic. It is able to overthrow the order of things and reconceive the world time and again.

This reading of scientific thinking as subversive, visionary, and evolutionary is quite different from the way science was understood by the positivist philosophers, but is also different from the fragmented, sometimes dry image of science provided by some more modern philosophical reflections on science. The aspect of science that I seek to illuminate in these pages is its critical and rebellious ability to reimagine the world again and again.

If this reimagining of the world is a central aspect of the scientific enterprise, then the beginning of this adventure is not to be sought in Newton's laws of motion, in Galileo's experiments, or Francis Bacon's reflections. Nor even in the early and mathematical constructions of Alexandrian astronomy. It must be sought in what can be called the first great scientific revolution in human history—Anaximander's revolution.

There is no doubt that Anaximander's importance in the history of thought has been underrated.* I believe that this has happened for several reasons. On the one hand, in the ancient world, his contributions were recognized by authors of a scientific bent, including Pliny (as quoted in the epigraph to this book), but Anaximander was generally seen by the ancients, including Aristotle, as the proponent of a naturalistic approach to knowledge that was fiercely opposed by other cultural currents and that had not yielded much in the way of results. The naturalistic project, indeed, had yet to bear the rich fruits it would bear with modern science, after a long process of maturation and numerous methodological adjustments.

At the root of today's underestimation of Anaximander's thought, on the other hand, lies the pernicious modern separation between science and the humanities.

*The situation is changing. Several recent studies converge on this point. Daniel Graham, in *Explaining the Cosmos: The Ionian Tradition of Scientific Philosophy*, comes to conclusions very similar to the ones in this book. In the introduction to the essay collection *Anaximander in Context*, by Dirk Couprie, Robert Hahn, and Gerard Naddaf, we read, "We are convinced that Anaximander was one of the greatest minds that had ever lived, and we felt that this had not been sufficiently reflected in the scholarship, until now." Couprie, who has studied Anaximander's cosmology in depth, concludes, "I do not hesitate to put him on a par with Newton."

I am aware that my mainly scientific training makes evaluating the contributions of a thinker who lived some twenty-six hundred years ago a risky proposition, but I am convinced that most if not all of today's assessments of Anaximander's contribution suffer from the inverse problem—the difficulty that specialists in history or philosophy have in evaluating the importance of insights whose nature and legacy are intimately scientific. It seems to me that even the authors quoted in the last footnote, who recognize without hesitation the greatness of Anaximander's contributions, fail to grasp the full extent of the historical importance of his multiple insights for the development of science. I seek to highlight that importance in these pages.

Therefore I examine Anaximander not as a historian or as an expert in Greek philosophy, but as a scientist of today keen to reflect on the nature of scientific thinking and its role in the long-term development of civilization. In contrast to the majority of texts about Anaximander, my goal is not to reconstruct as faithfully as possible his thought and conceptual universe. For this reconstruction, I rely on the painstaking, magisterial work of classicists and historians such as Charles Kahn, Marcel Conche, and, more recently, Dirk Couprie. My goal is not to challenge the conclusions of their reconstructions; it is to shed light on the profundity of the thought that emerges from them, and the role of Anaximander's insights in the development of universal knowledge.

A more subtle reason for the underestimation of Anaximander's thought and of Greek scientific speculation in general lies in what I believe is a common misunderstanding of certain central aspects of scientific thought.

Facile nineteenth-century certainties about science—in particular the glorification of science understood as definitive knowledge of the world—have collapsed. One of the forces responsible for their dismissal has been the twentieth-century revolution in physics, which led to the discovery that Newtonian physics, despite its immense effectiveness, is actually wrong, in a precise sense. Much of the subsequent philosophy of science can be read as an attempt to come to grips with this disillusionment. What is scientific knowledge if it can be wrong even when it is extremely effective?

A wide current in the philosophy of science has reacted by seeking to save a basis for certainty in science. Scientific theories, for example, have been interpreted as constructions whose value is limited to their directly verifiable consequences, within given domains of validity. The knowledge content of scientific theories has been interpreted as restricted to the ability to give predictions. In this way, in my opinion, we lose sight of the qualitative aspects of scientific knowledge and in particular of science's ability to subvert and widen our vision of the world. These qualitative aspects are not only inextricable from scientific thinking and essential for its functioning—they even constitute its primary motivation and reason of interest.

At the opposite end of the spectrum, another wide current of contemporary culture belittles scientific thinking and promotes widespread antiscience feelings. In the early twenty-first century, in many corners, rational science has come to be seen as suspect; forms of irrationalism have emerged in cultural circles and everyday life. Antiscientism feeds on the disillusionment over science's inability to deliver definitive visions of the world—on the fear of accepting ignorance. False certainties are preferred to lack of certainty.

But answers given by natural science are not credible because they are definitive; they are credible because they are the best we have now, at a given moment in the history of knowledge. Lack of certainty is anything but weakness. Instead, it constitutes—and has always constituted—the very strength of rational thinking, understood as curiosity, rebellion, and change. It is precisely by not taking its answers as definitive that science can continue to improve them.

From this point of view, three centuries of Newtonian science do not constitute Science. On the contrary, they are little more than a moment of rest along the way, in the shadow of a great success. In challenging Newton's theories, Einstein did not question the possibility to better discover how the world works. On the contrary, he followed in the footsteps of Maxwell, Newton, Copernicus, Ptolemy, Hipparchus, and Anaximander, all of whom advanced knowledge by challenging the received vision of the world, continuously improving it—recognizing errors and learning to look further and further ahead.

The advances achieved by these great scientists (and by innumerable other minor ones) have repeatedly changed not just our worldview but even the very rules of thinking that structure that worldview. I believe that looking for a key to unravel all problems—a methodological and philosophical fixed point to which this intellectual adventure could be anchored—is to betray science's very nature, which is intrinsically evolutionary and critical.

For some time now, humanity has discovered a path skirting the certainties of those who claim to know ultimate truths, while at the same time avoiding the downfall of claiming—as many claim today—that all truths are equal, each within its own cultural context, and we cannot distinguish true from false. This is the point of view that I shall seek to articulate in the final part of this text.

To look back at the ancient origin of scientific thinking, to the very first steps in the direction of rational inquiry about nature, is therefore here a way to shed light on some central aspects of the nature of this thought.

I think this reflection is important also for today's fundamental science. We are still immersed in the scientific revolution opened by Einstein.[1] To speak of Anaximander is also to grapple with the meaning of this revolution. My main scientific activity is in this field, and in particular in quantum gravity, a major open problem at the heart of today's theoretical physics. To address such a problem we likely need to change once again our understanding of the nature of time and space.[2] Anaximander succeeded in changing the old understanding of space, transforming the world from a closed box with the Heavens above and the Earth below to an open space in which Earth floats. I believe that only by understanding how such immense transformations of the world as Anaximander's are possible—and in what sense they are "correct"—can we hope to confront challenges like the changes in the notions of space and time demanded by the quantization of gravity.

Finally, there is a third thread running through this book: the discussion of a vast problem for which I can pose questions more than I can propose answers. As we examine the earliest ancient manifestations of rational thinking about nature, we are naturally led to examine the mode of knowledge that historically preceded it—a mode of knowledge that today still affirms itself as an alternative to rational thinking. This is the mode of knowledge from which rational thought was born and differentiated itself, and against which it rebelled and still rebels.

When he "opened the doors of nature" (in Pliny's words), Anaximander ignited a conflict between two profoundly different ways of thinking. On the one hand, there was the dominant mythical and religious way of thinking, based in large measure on the existence of certainties that, by their very nature, could not be called into question. On the other hand, there was the new way of looking at the world, based on curiosity, rejection of certainties, and change. This conflict has run through the history of Western civilization, century after century, with alternating outcomes. It is still open.

After a period in which these opposing modes of thinking seemed to have coexisted peacefully, the clash appears to be reemerging today. Numerous voices, from political and cultural viewpoints that otherwise diverge greatly, once again speak out on behalf of irrationality and the primacy of religious thought. This renewal of the clash between positive and mythico-religious thought takes us back to the conflicts of the Enlightenment. But I think that it is a mistake to consider only the past decade or the past few centuries in attempting to clarify terms. The clash is more profound. It is measured in millennia rather than centuries, for reasons relating to the slow evolution of human civilization, the deep structure of its conceptual organization, and its gradual political and social evolution.

These are vast themes, and I can do little more than

ask questions and seek out some grounds for reflection in the final chapters of the book; but I believe that these themes are central to our world and its future. Every day, the uncertain outcomes of this conflict shape the lives and fate of all humanity.

I do not wish to overstate the importance of Anaximander. In the end, we know very little about him. But twenty-six centuries ago, on the Ionian coast, somebody opened a new path to knowledge and a new route for humanity. A thick fog veils the sixth century before the Common Era, and we know too little of the man Anaximander to be able to attribute this immense revolution to him with certainty. Still, this revolution, the birth of a thinking based on curiosity and change, took place. In the end, whether this change was wrought personally by Anaximander, or whether "Anaximander" is simply the name used in ancient sources to identify it, matters little.

This extraordinary revolution, begun twenty-six centuries ago on the coast of present-day Turkey, and in which we are still immersed, is the topic of this book.

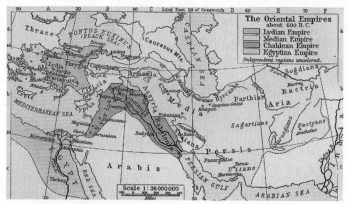

Figure 2. A nineteenth-century map showing the empires of the
Middle East around 600 BCE.

ONE

THE SIXTH CENTURY

The sixth century before the Common Era (BCE) is not among the most widely familiar historical periods. When Anaximander was born in Miletus in 610 BCE, the Golden Age of Greek civilization, the time of Pericles and Plato, was still nearly two hundred years in the future. Tarquin the Elder, according to tradition, reigned in Rome. At around the same time, the Celts founded Milan, and Greek settlers from Anaximander's Ionia founded Marseille. Homer (or whoever for him) had composed the *Iliad* two centuries earlier, and Hesiod had already composed the *Works and Days*, but none of the other Greeks' illustrious poets, philosophers, and dramatists had begun writing. Sappho, still a girl, was living on an island near Miletus.

In Athens, whose power was just beginning to grow, Draco's strict code of law ruled, but Solon, who would write the first constitution to incorporate democratic elements, had already been born.

The Mediterranean world was far from primitive. Humans had been living in cities for at least ten thousand years. The great Kingdom of Egypt had been in existence for some twenty-six centuries—the same span of time that separates Anaximander from us.

Two years before Anaximander's birth, the city of Nineveh had fallen, a momentous event that marked the end of Assyria's vast and brutal power. Babylon, with more than two hundred thousand inhabitants, was once again the largest city in the world, as it had been for thousands of years. Nabopolassar ruled over Babylon, but the city's return to splendor would be short-lived. Under Cyrus I, Persia's power was already stirring in the east, and the Persian Empire would soon take control of Mesopotamia. In Egypt, it was the last year of the long reign of the great Psamtik I, the first pharaoh of the Twenty-sixth Dynasty, who had won Egypt's independence back from the dying Assyrian Empire and restored prosperity to his realm. Psamtik I had established close relations with the Greek world: he had enrolled numerous Greek mercenaries in his army and encouraged Greeks to settle in Egypt. Miletus maintained a flourishing port of call in Egypt, at Naucratis. Anaximander, then, likely had abundant first-hand information about Egyptian culture.

Josiah of the House of David reigned over Jerusalem. With the Assyrian Empire weakened and Babylonia not yet restored to full power, he took advantage of international instability to reaffirm Jerusalem's pride by imposing exclusive worship of the single God, Yahweh. He destroyed all the ritual objects of other gods (such as Baal and Astarte), tore down their temples, massacred their living priests, and exhumed and burned upon their altars the bones of the dead priests,[1] establishing a mode of behavior toward other religions that would later characterize triumphant monotheism. Before Anaximander's death, the Israelites fell captive again and were deported to Babylon, where they once again knew servitude—a servitude from which they would once again win their freedom, as they had centuries before from Egypt, with Moses.

Echoes of these events surely reached Miletus. News of happenings elsewhere in the world probably did not. Northern Europe was passing from the Bronze Age into the Iron Age. In the Americas, the ancient Olmec civilization was already waning. In northwest India, the Mahajanapada kingdoms had formed. Mahavira, a contemporary of Anaximander in India, founded Jainism and preached nonviolence toward all living beings. Already the Indoeuropeans of the West were focusing on how to better think about the world, while those of the East reflected on how to better live.

In China, King Kuang of Zhou had recently ascended to the throne as the twelfth emperor of the illustrious Zhou Dynasty. It was the so-called Spring and Autumn Period, a time of decentralization of power and feudal battles—and of cultural diversity and creativity as well. China would not know a similar culturally productive era for a long time to come. This has perhaps been the price paid for an internal stability that, while imperfect, has nevertheless far exceeded that of the ferocious West, endlessly at war.

Human civilization, thus, had been in existence and highly structured for thousands of years when Anaximander was born at the dawn of the sixth century BCE. The traffic of goods and ideas among continents was intense. At Miletus, it was perhaps already possible to purchase Chinese silk, as would be the case two centuries later in Athens. Most men survived by working the land, raising animals, fishing, hunting, or engaging in trade; others, just like today, amassed power and riches by going to war against one another.

KNOWLEDGE AND ASTRONOMY

What was the state of knowledge and the cultural climate of Anaximander's world? It is hard to say, because the sixth century, in contrast to the wordy epochs to follow, left relatively few written documents. Some great works whose influences endure to this day had already been compiled, including major parts of the Bible (Deuteronomy probably dates from around this time); the Egyptian *Book of the Dead*; and the great epics such as *Gilgamesh*, the *Mahabharata*, the *Iliad*, and the *Odyssey*—splendid, grandiose stories in which humanity reflects upon itself, its dreams, and its follies.

Writing had begun three thousand years earlier. Written laws had been in existence for at least twelve hundred years, since Hammurabi, sixth king of Babylon, had his inscribed upon splendid basalt steles that were displayed in every city of his vast empire. One of these steles is now on display at the Louvre, in Paris, and it is difficult to remain unmoved when viewing it and reading the translation of its text.

And scientific knowledge? In Egypt and especially in Babylon, basic principles of mathematics had been developed, known to us thanks to the unearthing of collections of mathematical problems and answers. Young Egyptian scribes learned to perform simple division so that sacks of grain could be distributed equally among creditors or according to certain proportions. (One merchant has twenty sacks of grain with which to pay two workers, one of whom worked triple the time of the other. How many sacks must he give each?)

Calculation techniques were known for dividing numbers by two, three, four, and five, but not by seven. A problem that involved division by seven had to be reformulated in other terms. The constant now known as π (namely 3.14 . . .) was used as today to calculate the perimeter of a circle based on its diameter but a precise value for π was unknown.

The Egyptians knew that a triangle with sides in the proportion 3:4:5 has a right angle. Egyptian and Babylonian mathematical knowledge was roughly equivalent to that of an advanced student in today's second or third grade. One often reads of the extraordinary development of ancient Babylonian mathematics. This is correct, but it must be interpreted properly: it means that the Babylonians developed the concepts that, in our time, are studied by seven-year-olds. The point is that it has been anything but easy for humanity to collect the knowledge that today's children learn in elementary school.

Whether in Egypt, Babylonia, or Jerusalem, in Crete or Mycenae, in China or Mexico, knowledge was concentrated at royal and imperial courts. The fundamental form of human political organization in the first great civilizations was the monarchy, a centralized form of power. I think we can say that the great monarchies *were* the great civilizations. Laws, trade, writing, knowledge, learning, religion, political structure—all of these things existed mainly within the royal and imperial palaces. The structure of monarchy allowed civilization to develop. It guaranteed the security and the stability needed for the complexity of civilization. Stability that sometimes held and sometimes failed, as it does today.

Figure 3. Cuneiform tablet inscribed in Nineveh in the seventh century BCE. It records observations of the position of the planet Venus made one thousand years earlier, during the reign of Ammisaduqa. (*British Museum*)

The Babylonian court maintained registers of important and noteworthy facts. These included the price of grain, descriptions of catastrophic events, and—crucial for the future development of science—records of astronomical data such as eclipses and planetary positions. Eight centuries later, during the Roman Empire, Ptolemy still considered data from Babylon's ancient archives reliable enough to use. He complains about not having access to the complete Babylonian documents on planetary positions but avails himself of eclipse tables compiled during the reign of Nabonassar around 747 BCE, a century before Anaximander. Ptolemy even uses the start of Nabonassar's reign as year zero for his elaborate astronomical calculations.

The recording of astronomical data is even more ancient. We have a cuneiform tablet, shown in figure 3, indicating the (correct) positions of Venus in the Heavens, over the course of many years, during the reign of Ammisaduqa, namely around 1600 BCE, a millennium before Anaximander.

It is useful to stop and reflect on this ancient astronomy because it relates to future science. What did these data mean to Babylonians? Why did they record them? Why did Babylonians observe the Heavens?

The reason for the Babylonian interest in the Heavens is clearly written on the hundreds of thousands of ancient tablets still in existence. On the one hand, human beings took note of the patterns of certain astronomical phenomena and made practical use of them. On the other hand, they sought to establish a relationship between astronomical and earthly phenomena. Let me discuss these two motivations separately:

Mediterranean climate requires farmers to carefully follow annual cycles. How to follow these rhythms in a world without calendars or newspapers, and in a region where seasons are not clearly marked? The sky and the stars offer a relatively simple solution. Human beings had known this for centuries, and the knowledge was widespread. Hesiod, for example, in the *Works and Days*, describes these phenomena in beautiful language:

> When . . . the star Arcturus leaves the holy stream of Ocean and first rises brilliant at dusk, after him the shrilly wailing daughter of Pandion, the swallow, appears to men when spring is just beginning. Before she comes, prune the vines, for it is best so.

And,

> But when Orion and Sirius are come into mid-heaven, and rosy-fingered Dawn sees Arcturus, then cut off all the grape-clusters, Perses, and bring them home. Show them to the sun ten days and ten nights: then cover them over for five, and on the sixth day draw off into vessels the gifts of joyful Dionysus. But when the Pleiades and Hyades and strong Orion begin to set, then remember to plough in season: and so the completed year will fitly pass beneath the earth.

Perses, the poem's addressee, was Hesiod's brother. Again:

> But if desire for uncomfortable sea-faring seize you; when the Pleiades plunge into the misty sea to escape Orion's rude strength, then truly gales of all kinds rage.[2]

In short, according to Hesiod, it is easy to determine the month of the year by observing the stars: when Arcturus appears over the sea in the evening, it is spring; autumn begins when the constellation Orion and Sirius, the Dog Star, are overhead; and autumn ends and winter begins when the Pleiades have fully set. According to the

Bible, too, Yahweh created stars on the fourth day to "be for signs, and for seasons, and for days, and years."[3]

It is clear then that the movement of the sun and stars in the sky had been understood by peasants for centuries, with a clarity far superior than an average educated person of today. Hesiod considers it common knowledge that to determine the time of year, one need simply glance at the constellation visible in the east at dawn. Few university professors could do so today.

At times, Hesiod seems to attribute to the stars themselves the *causes* of human perception, as in his extraordinary verses on the heat of summer:

> When the thistle blooms and the chirping cicada sits on trees and pours down shrill song from frenziedly quivering wings in the toilsome summer, then goats are fatter than ever and wine is at its best; women's lust knows no bounds and men are all dried up, because the dog star parches their heads and knees and the heat sears their skin.

I find it hard to tell whether this attribution of men's feebleness to Sirius is to be taken literally, or if "the dog star" here simply indicates summer itself. Such a distinction may be irrelevant in this context: Hesiod may simply mean that when Sirius is high in the sky (in summer), then men are feeble, without being concerned with causal relations. We ourselves sometimes say, "The beginning of the afternoon makes me sleepy," without stopping to ponder whether the fatigue is caused by the time of day or the noonday meal.

This brings me to the second and more important role of ancient astronomy: the effort to establish a link between celestial phenomena and events of direct importance to human beings. The connection between heavenly and human events has been a matter of concern since earliest antiquity—whether the connection was perceived as causal or a matter of temporal coincidence, if such a distinction was even meaningful in the sixth century BCE. Returning to Babylonia, we read, for example, in the tablet of figure 3: "On the fifteenth day of the month, Venus disappeared.* It was gone from the Heavens for three days. Then on the eighteenth day of the eleventh month, it reappeared in the east. New springs began to flow, the god Adad sent rain, the goddess Ea sent her floods."[4]

This type of presentation, linking events in the Heavens and on Earth, is nearly universal in the ancient cuneiform texts about astronomy that have come down to us. Here is another example from the same period, interpreting the appearance of the Sun in the sky before dawn:

> If, in the month of Nisanu [the first month of the Babylonian calendar, roughly March-April], the Sun at dawn appears splattered with blood and the light is cold, then rebellion in the land shall not die and the god Adad shall accomplish slaughters.
>
> If, in the month of Nisanu, the dawn appears splattered with blood, there shall be battles in the land.
>
> If, on the first day of the month of Nisanu, the dawn appears splattered with blood, there shall be great harshness and human flesh shall be consumed.
>
> If, on the first day of the month of Nisanu, the dawn appears splattered with blood and the light is cold, the king shall die, and there shall be mourning in the land.

*A reminder to reader: Venus sometimes appears in the sky in the east, sometimes in the west, and occasionally vanishes altogether.

If this shall happen on the second day of the month of Nisanu, one of the king's high officials shall die, and mourning in the land shall continue.

If on the third day of the month of Nisanu, the dawn shall appear splattered with blood, there shall be an eclipse.

All Babylonian documents show clearly that the gathering of astronomical data such as eclipses and planetary positions was motivated by the belief that this information was connected to events of direct interest to humanity such as wars, floods, the deaths of leaders, and so forth.

This belief is utterly mistaken, of course, but to this day it is shared by perhaps the majority of the people—even in highly educated nations—including some in positions of very high authority.

In Babylonia, then, men gathered astronomical data and sought patterns and relations between happenings in the Heavens and those on Earth. Relations among celestial events were of concern as well. We cannot exclude the possibility that someone in Anaximander's time may have been able to predict eclipses with some degree of accuracy—or, in the case of solar eclipses, at least predict when there might be an eclipse. This is not difficult given the existence of patterns in the time series of the eclipses. An intelligent and motivated person can find these patterns rather easily by examining the data.*

---

*Every eighteen years, eleven days, and eight hours, the Sun, the Moon, and Earth return to roughly the same relative positions. The time series of eclipses repeats almost identically after this period, called the Saros cycle, making approximate predictions relatively simple.

The ancient Greeks report with marvel that Thales, Anaximander's teacher, predicted a solar eclipse, though no one knew how he managed to do so. We do not know if this story is reliable, but in all likelihood, Thales had traveled to the Babylonian court.

Developments on the other side of the planet illustrate another purpose of astronomy. In the sixth century BCE in China, the famous imperial institute of astronomy had probably already been established. According to *Shu Jing* (*The Book of Documents*), which may date from around 400 BCE, the study of astronomy in China began under the legendary Emperor Yao, who lived some two thousand years before the Common Era. The *Shu Jing* relates that Yao

> commanded Xi and He, in reverent accordance with their observation of the wide heavens, to calculate and delineate the movements and appearances of the sun, the moon, the stars, and the zodiacal spaces, and so to deliver respectfully the seasons to the people.

Xi and He both had two sons who were sent to the four corners of the Earth, each with the mission to determine solstices and equinoxes. Finally, the emperor turned once again to Xi and He:

> Ah! you, Xi and He, a round year consists of three hundred, sixty, and six days. By means of an intercalary month do you fix the four seasons, and complete the determination of the year. Thereafter, in exact accordance with this, regulating the various officers, all the works of the year will be fully performed.[5]

The main issue driving the investigation of astronomical phenomena and the foundation of the institute is

clearly the problem of the calendar.* The development
of real astronomical knowledge in China dates more
probably from the Han period, some two centuries after
Anaximander and much later than the parallel developments in Babylonia.

Over the course of the following millennia, Chinese
astronomers developed rudimentary techniques for predicting planetary positions and eclipses. Still, though the
imperial institute of astronomy in China existed almost
without interruption for more than twenty centuries—
and though it had at its disposal astronomical observations gathered over thousands of years, along with some
of the empire's most brilliant minds, chosen on the basis

---

*The problem of the calendar vexed all civilizations, from the Maya to
the Chinese, from Julius Caesar to Pope Gregory. The easiest way to
keep track of days is to count lunations and to determine the day by
observing the lunar phase. Full and new moons, first and second quar-
ter moons, are easy to identify; thus, between one phase and the next,
one need only count the days (roughly seven, namely a week). The sim-
plest and most widely followed calendar, hence, is the lunar calendar.
However, it presents two problems: First, for agriculture and the track-
ing of long periods, it is the annual solar cycle that is relevant. In contrast to lunar cycles, though, it is difficult to mark the beginning and
end of the solar cycle (thus Emperor Yao's need for specialists to determine solstices and equinoxes). Second, a month does not contain an
exact number of days, just as a year does not contain an exact number of
months or days. This creates a need for months with more days and
months with fewer days to stay synchronized with the lunar cycle—and,
furthermore, it becomes impossible to keep months and years in align-
ment while respecting lunar and solar phases. The solution adopted by
the modern world—with months that vary in length, not synchronized
to the Moon's motion; leap years every fourth year (minus one every
hundred years but plus one every four hundred years); and days of the
week independent from the date—is extremely complicated and seems
reasonable only to those accustomed to it. Other cultures have come up
with other solutions, all equally intricate.

of merit following a series of rigorous exams—its results were far from brilliant. Just three hundred years ago, the institute's ability to predict celestial phenomena was vastly inferior to that of Ptolemy's *Almagest*, written more than fifteen hundred years earlier. What's more, the institute had not yet grasped the fact that the Earth is round.

Beyond the calendar, Chinese imperial authorities' interest in astronomy was driven by religious and ideological considerations. The official Confucian cult, like religions in Greece and modern Europe, taught that Heaven is the seat of the divinity. The emperor was the intermediary between Heaven and Earth, the guarantor and executor of an order that was at once worldly, social, and cosmic. For Confucius, this function was carried out by means of rites—just as, in the Catholic Church, the ritual of Mass renews and strengthens the alliance between God and humanity and reestablishes order for human beings lost in the confusion of everyday reality. The imperial institute of astronomy had the crucial mission of establishing the official timetable for rites, "to establish concord with the august Heaven."

With all due caution, certain analogies between Chinese and Babylonian astronomy can help us shed some light on Babylonia's interest in the stars, and its relation to the later Greek one. What Chinese astronomy teaches us is that observation of celestial phenomena over many centuries, with the full support of political authorities, is not sufficient to lead to modern science (that of Copernicus, Kepler, Galileo, and Newton). What's more, it does not even lead to the development of an effective, predictive, precise mathematical theory (that of Hipparcos and Ptolemy); nor does it lead to clear advances in understanding the structure of the world (like Anaximander's). Similarly, ancient Mesopotamian civilizations observed celestial phenomena in a continuous and sustained manner for millennia, but they never advanced beyond the gathering of fairly imprecise data, interpreted within an utterly mistaken conceptual framework that linked them to earthly events.*

Chinese motivations and the Chinese spirit in studying the Heavens were not necessarily the same as the ones that inspired the work of Babylonian astronomers. There were differences between China and Babylonia. But what the two had in common was to show that it was possible to study astronomy in the context of a world-view utterly foreign in form and motivation to that of Anaximander, Ptolemy, Copernicus, and Einstein. What made the motivations and the logic of ancient (pre-sixth century BCE) Middle Eastern and ancient (pre-seventeenth-century CE) Chinese astronomy so different from those of Anaximander, Ptolemy, Copernicus, and Einstein is the central subject of this book.

But let us return to the region that concerns us: the Greek world in which Anaximander was born.

THE GODS

Hesiod, who wrote a century before Anaximander's birth and must have been well known in Anaximander's Miletus, gives us a general idea of Greece's cultural world before Anaximander. Hesiod's world is deeply human, built upon the rigor of peasant labor and positive, wholesome moral values. Hesiod considers questions about the meaning of human life and work (the

---

*Later in the book I discuss in greater detail the precise meaning of the word "mistaken" in this context, with regard to an awareness of the cultural relativity of truths and values.

Works and Days) and about the birth and history of the
universe (the *Theogony*), foreshadowing themes and ideas
that would drive the great philosophical inquiries of
future centuries.

Hesiod's answers to these questions are complex, to be
sure, but are clearly made of the same stuff that we find
all over the world, in particular in the Tigris and
Euphrates Valley—the stuff of gods and myths.

Consider one example: How did the world come into
being? What is the world made of? Hesiod's answer
comes near the beginning of the *Theogony*:

> Verily at the first Chaos came to be, but next wide-
> bosomed Earth, the ever-sure foundations of all the
> deathless ones who hold the peaks of snowy Olympus,
> and dim Tartarus in the depth of the wide-pathed Earth,
> and Eros, fairest among the deathless gods, who
> unnerves the limbs and overcomes the mind and wise
> counsels of all gods and all men within them. . . .
>
> And Earth first bare starry Heaven, equal to herself, to
> cover her on every side, and to be an ever-sure abiding-
> place for the blessed gods. And she brought forth long
> Hills, graceful haunts of the goddess-Nymphs who dwell
> amongst the glens of the hills. She bare also the fruitless
> deep with his raging swell, Pontus, without sweet union
> of love.
>
> But afterwards she lay with Heaven and bare deep-
> swirling Oceanus, Coeus and Crius and Hyperion and
> Iapetus, Theia and Rhea, Themis and Mnemosyne and
> gold-crowned Phoebe and lovely Tethys. After them was
> born Cronos the wily, youngest and most terrible of her
> children, and he hated his lusty sire. . . .

And so on in this splendid vein. Hesiod's account of
the world's origins and structure is very similar to those
found in all other civilizations. Following is the story of

the creation of the world given in the *Enuma Elish*, which was recited on the fourth day of the new year in Babylonia. The text was discovered inscribed upon cuneiform tablets of the twelfth century BCE (half a millennium before Hesiod) in Ashrupanipal's palace in Nineveh:

> When in the height heaven was not named,
> And the earth beneath did not yet bear a name,
> And the primeval Apsu, who begat them,
> And Mummu-Tiamat, the mother of them both
> Their waters were mingled together like a single body,
> And no hut was formed, no marsh was to be seen;
> When of the gods none had been called into being,
> And none bore a name, and no destinies were ordained;
> Then were created the gods in the midst of heaven.
> Lahmu and Lahamu were called into being, and were called by name.
> Before they had grown in age and stature Ansar and Kisar were created, who surpassed the others.
> Long were the days, then there came forth Anu, their heir, rival of his father.[6]

And so on for hundreds of verses. The consonance with Hesiod's text is obvious. All of the evidence that has come down to us indicates that it is through these myths that humanity sought to give order to the world. Human beings interpreted earthly events as shaped by the powers of the gods and supernatural beings.

Stories of gods make up almost the entirety of ancient texts. Gods give structure to the world, appear as characters in all the great tales, justify the power of monarchies, are identified with that power, are invoked in judgments by individuals and groups, and serve as guarantors of law.

All ancient civilizations have in common this centrality of the divine. Gods played an absolutely fundamental role in civilization for at least as long as written evidence survives.*

Why did all human beings create and share a system of thought in which gods play such an overwhelming role? When and why did this strange structure of thought come into being? These questions are basic to understanding the nature of civilization, and the answers are still far from clear. But the centrality and universality of polytheistic gods as a basic element in the structure of ancient thought are beyond question.[7] When Anaximander was born, the foundations of knowledge were sought entirely in myth and divinity.

MILETUS

An atmosphere very different from that of Jerusalem, Babylonia, and Egypt blew in the young cities of Greek's emerging civilization, in strong geographical, economic, commercial, and political expansion. The diversity already manifested itself in all forms of expression of this young culture—Ionic sculpture, for example, which anticipated the naturalism and variety of classical Greek art (see figure 4, p.21).

The novelty of this culture shone even more clearly in the earliest lyric poetry, of a lacerating newness with respect to anything that had been written to that point:

> Seems to me the equal of the gods,
> he, who sits in your presence and hears near him
> your sweet voice and lovely laughter;
> my heart beats fast in my bosom.

---

*In Hammurabi's Code, mentioned earlier, the text is introduced by Hammurabi, who claims that the law was dictated to him by the god Marduk, just as Jewish law was said to have been dictated to Moses by Yahweh.

I see you even a little and I cannot speak anymore,
my tongue is useless
at once a subtle fire races under my skin,
my eyes see nothing,
my ears ring,
I sweat
all my body is seized with trembling.
I am paler than grass and in my madness
I seem close to be dead . . .[8]

Marvelous.

But it was the new complexity of the political structure that especially reflected the radical novelty of the
Greek world. While peoples around the planet struggled
to attain stability by means of great kingdoms and
empires, modeling themselves upon the millennial reign
of the pharaohs, Greece remained divided into cities
fiercely jealous of their independence. This fragmentary
structure turned out to be anything but a source of weakness—it was at the heart of the extraordinary dynamism
that drove the immense success, political as well as cultural, of the Greek world.*

Anaximander's intelligence did not blossom in the rich
and efficient bureaucracy of the pharaoh's scribes, nor in
the highly structured court of ancient Babylonia, where
the knowledge of the ancient world was stored, but in a
young, independent, flourishing Ionian city on the sea:
merchant ships came and went, and the citizens of
Miletus probably thought of themselves as masters of

---

*This pattern probably repeated itself in late medieval and early mod-
ern Europe. While other world cultures were completing a process of
political unification and imperial stability, the failure of this process in
Europe brought about a different rate of growth that ultimately determined Europe's military, cultural, and political success.

their personal and civic destiny to a far greater degree than any pharaoh's subject.

Ionia was a small region on the coast of Asia Minor, made up of a dozen cities overlooking the sea and protected by a steep, jagged rocky coast. Here, in this small strip of land, obscure and of minor importance in world history, the earliest example of critical thinking came into being—the free spirit of investigation that would come to define Greek and, eventually, modern thought. Human civilization owes an enormous debt to this land—greater, perhaps, than what it owes to Egypt, Babylonia, and Athens.[9]

In the interior of the Ionian coast, in Asia Minor, was the rich kingdom of Lydia, which a few decades earlier had minted the world's first coins. Alyattes II, king of Lydia, took the throne in 610 BCE, the year Anaximander was born, and pursued the war against Miletus waged by his father, Sadyattes. Soon, though, Alyattes was forced to turn his attention to hostilities against Babylonia and the Mede Kingdom, pressing from the southeast. He made peace with Miletus and left the city undisturbed. The tomb of Alyattes still stands in the plain between Lake Gigea and the river Hermus, to the north of Sardis in modern Turkey. It is a large mound of earth covering a structure of enormous boulders. Large stone phalluses still stand on its summit.

The cities of Ionia were inhabited by Greeks. They had probably arrived from various parts of Greece long before, perhaps a century or two after the Trojan War, and mingled with the existing local populations. The cities of Ionia were independent but joined in a confederation, the Ionian League, mostly cultural and religious in nature.¹⁰ The league delegates gathered at Panionium, a sanctuary dedicated to Poseidon Helikonios. In 2005, the probable remains of this sanctuary were identified on the slopes of the peninsula of Mount Mycale, south of modern İzmir. Something of a Greek outpost into the ancient civilizations to the south, Ionia was renowned for its wealth and fertility.

In addition to prized local products, such as olive oil from the groves that to this day surround the ruins of Miletus, Ionia's wealth derived from trade with peoples of the north, near the Black Sea. Ionia controlled the transit route that centuries earlier had made Troy rich—and that Greeks had paid dearly to control. Trade with Asia was important, too, via the caravan

Figure 4. Anavyssos Kouros, life-size statue in marble, probably sculpted during Anaximander's lifetime. (*National Archaeological Museum of Athens*)

routes that cut across Asia Minor and arrived at Syria's markets. The ships of the Phoenicians, from whom the Greeks took their alphabet, arrived from the south. Ionia was the hinge between West and East.

Greek cities had a substantial number of slaves; a mixed economy (comprising agriculture, crafts, and trade); and free citizens who took up arms when necessary. In the sixth century, Miletus was the most prosperous of the Ionian cities and probably of the entire Greek world (the power of Athens and Sparta developed later), and it was the closest city to the great civilizations to the south. Herodotus the historian called Miletus "the jewel of Ionia."[11]

Miletus had been in existence long before Greeks colonized it. It is mentioned as "Millawanda" in the Hittite annals of Mursili II, which record that the city allied itself with Uhha-Ziti's rebellion in Arzawa in 1320 BCE. As a result, Mursili ordered his generals Mala-Ziti and Gulla to raze Millawanda. Modern archaeologists have discovered evidence of this destruction. The city was then fortified by the Hittites, probably to defend it against Greek attackers, but subsequently destroyed several times more by various invaders.

Herodotus relates that Neleus, the youngest son of Codrus, king of Athens, founded Greek Miletus around 1050 BCE. Neleus and his troops killed the native men and took their women as wives. But the monarchy in Miletus died out by the end of the eighth century, following a dispute between two descendants of the royal house of Neleus, Amphitryon and Leodamas. Amphitryon had Leodamas killed and took power by force. Leodamas's exiled son then returned with a band of followers who fought against and killed Amphitryon. Soon after, peace was restored, but the monarchy had lost authority. Citizens elected a legislator and a "temporary dictator," Epimenus. The city was then governed by the *prytaneion*, an elected oligarchical council of magistrates, which repeatedly degenerated into tyranny.

Miletus therefore experienced a complex political process that recalls that of Athens and the later familiar saga of Rome: a king driven away by the aristocracy, which in turn finds itself threatened by a rich merchant class, which in turn plays the role of mediator between the aristocracy and the world of craftsmen and farmers.

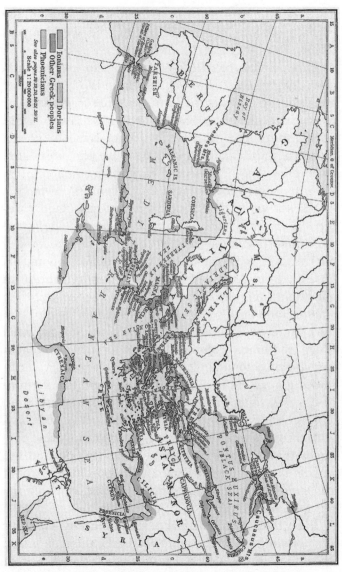

Figure 5. The expansion of Greece and Phoenicia in the midsixth century.

Long political battles occurred, characterized by the conflict between *Ploutîs* (Πλουτις), the wealthy, and *Cheiromáche* (Χειρομαχα), the workers.

This political complexity was the characteristic that most distinguished Greek culture from the kingdoms of the East, and it was at the heart of the emerging cultural revolution. In 630 BCE, twenty years before Anaximander, the dictator Trasibulus came to power, most likely with the support of the people, against the aristocracy. Trasibulus would play an important role in the city's history, leading it to the peak of its power.

Miletus was a flourishing city when Anaximander was born in the early sixth century BCE. It was one of the most important commercial ports in the Greek world, if not the most important one, as well as the most populous Greek city in Asia, with some one hundred thousand inhabitants. It controlled a small but important maritime empire comprising several dozen colonies, most of which were scattered along the coast of the Black Sea. Pliny the Elder lists ninety colonies founded by Miletus. There were Ionian colonies in Italy and today's France. The city traded grain from its Scythian colonies (in the territory of Ukraine) along with timber, dried and salted fish, iron, lead, silver, gold, wool, linen, ocher, salt, spices, and skins. From Naucratis came salt, papyrus, ivory, and perfumes that arrived in caravans from Ethiopia and the Middle East. Miletus produced and exported terra-cotta, arms, oil, furniture, textiles, fish, figs, and wine. Its fabrics were renowned.

The Milesian commercial port of Naucratis in Egypt was probably founded around 620 BCE, a decade before Anaximander was born. There was surely no lack of commercial and cultural contact with Egypt's ancient civilization. Egypt exerted a markedly strong influence on architecture: the first monumental Greek temples date from this period and were directly inspired by Egyptian architecture, both technically and stylistically.[12]

Colonies and trade generated not just wealth but also interactions with different peoples, ideas, and opinions. Miletus had economic and cultural ties to all of the Mediterranean and the Middle East. As its economy expanded, so did its worldview.[13]

Miletus, then, was wealthy, free, able on its own to fend off ambitious Lydia, and probably the Greek city most exposed to the advanced cultures to the south. Unlike the great cities of Mesopotamia and Egypt, Miletus had neither a grand royal palace nor a powerful priestly caste. It was a city of free citizens at the heart of a cosmopolitan culture, prosperous and in a phase of extraordinary artistic, political, and cultural flowering. Miletus, in short, was the heart of a first, flourishing humanism.

The beautiful ruins at Miletus today do not date back to Anaximander's time. The oldest ones are from the century after he lived. In 546, the year before he died, Anaximander would see Miletus vanquished by the Persian Empire, which was expanding in the void left by the fall of the Assyrian Empire. A few decades later, in 494, following an ill-fated attempt at revolt, Miletus would be sacked and razed by the Persians, who captured and enslaved most of its inhabitants and deported them to the Persian Gulf. The episode marked the end of Miletus's cultural primacy in ancient Greece.

By the middle of the fifth century, Miletus rose again and was rebuilt according to plans by Hippodamus, the great architect and revolutionary father of urban plan-

Figure 6, top. The theater of Miletus. Figure 7, below. The Market Gate of Miletus at the Pergamon Museum in Berlin.

ning. Some of the ruins in existence today date from this era, including the splendid theater later enlarged during the Roman period (figure 6).

The famous Market Gate of Miletus (figure 7), transported to Berlin's Pergamon Museum in 1907 and reconstructed there in 1928, dates entirely from the Roman period and bears witness to the enduring prosperity of Miletus under the Roman Empire.

Anaximander was probably an important citizen of Miletus. According to one source, Aelianus, he was the head of the Milesian colony in Amphipolis.[14] Thales, one of the Seven Sages of ancient Greek tradition, lived in Miletus shortly before Anaximander. Thales traveled

Figure 8. A cup crafted in Sparta in the sixth century attributed to the Arkesilas Painter. Some authors have seen in it the influence of Anaximander's ideas, with Earth in the form of a column and the Heavens (borne by Atlas) surrounding the Earth. The other character is Prometheus. (*Vatican Museums*)

widely and took part in civic affairs. They surely knew each other, and I discuss their relationship in chapter 6.

Ancient sources tell of a voyage by Anaximander to Sparta, where he reportedly built a sundial. According to Cicero, Anaximander saved the lives of many Spartans by predicting an earthquake. The story seems improbable, but one way or the other, accounts that place him in Amphipolis and Sparta depict Anaximander as a traveler and a respected and renowned figure. Some authors believe that he may have journeyed to Egypt via Naucratis.[15]

No descriptions of Anaximander have come down to us—only a mention of unknown reliability by Diogenes Laertius. He relates that Empedocles sought to emulate Anaximander by affecting solemn airs and a theatrical demeanor. Given that he committed his reflections to writing, Anaximander must have had access to written texts. Though, we know almost nothing of his readings, life, character, appearance, or voyages.

But it is Anaximander's ideas that interest us. These are outlined in the next chapter.

TWO

ANAXIMANDER'S CONTRIBUTIONS

If we knew history, we would find a great intelligence at the origin of any innovation.
—Émile Mâle, *L'art religieux de la XII siècle en France*

Anaximander wrote a treatise in prose, *On Nature* (Περὶ φύσεως), now lost. Only one fragment remains, quoted by Simplicius of Cilicia in his commentary on Aristotle's *Physics*:

ἐξ ὧν δὲ ἡ γένεσίς ἐστι τοῖς οὖσι, καὶ τὴν φθορὰν εἰς ταῦτα κατὰ τὸ χρεών
διδόναι γὰρ αὐτὰ δίκην καὶ τίσιν ἀλλήλοις τῆς ἀδικίας κατὰ τὴν τοῦ χρόνου τάξιν

The translation is disputed but might read as follows:

All things originate from one another, and vanish into one another
According to necessity;
They give to each other justice and recompense for injustice
In conformity with the order of Time.

Much has been written about this handful of obscure words, which can easily inspire fanciful interpretations. It is always difficult to interpret a passage out of its context with any degree of certainty. It is not this fragment of direct evidence that tells us what is interesting in Anaximander's ideas.

Instead, many Greek sources relay the content of Anaximander's book. To be sure, many of these sources are late, indirect, and occasionally unreliable. Among the most interesting is Aristotle, who writes only two centuries after Anaximander and discusses his ideas repeatedly. It is very likely that Aristotle had a copy of Anaximander's text in his celebrated library. The history of philosophy by Theophrastus, Artistotle's pupil and his follower in the direction of the peripatetic school, discusses Anaximander's ideas in detail. Theophrastus's work is lost, too, but a number of later sources contain abundant information about it—Simplicius, for example, who lived in Alexandria and Athens in the sixth century of the Common Era. More than one thousand years separate Simplicius from Anaximander.

The modern effort to reconstruct Anaximander's ideas from the numerous, motley, and late sources is thus a complex puzzle. To be sure, charred scrolls found during archaeological digs in ancient Roman libraries are now being unrolled and deciphered using ever more refined techniques. The same is true for x-ray readings of Egyptian mummies' bandages, which were often made from strips of papyrus manuscripts. So we can still hope that Theophrastus's text—or perhaps even Anaximander's—might come to light. (This is a real possibility. Anaximander's name appears in a fragment of a catalog listing authors whose works were in a Roman library recently unearthed in Taormina.[1]) But until then, to know the content of Anaximander's book we must rely on the reconstruction from the indirect sources.

Here I do not enter at all into the complex art of textual reconstruction. I summarize the main ideas that we can reasonably attribute to Anaximander relying on the reconstructions of the content of his book that appear to be most reliable to me. I follow in particular Charles Kahn, Marcel Conche, Dirk Couprie, and Daniel Graham, and I take a middle position between full severity—attributing to Anaximander only those ideas that can be traced back to him with absolute certainty—and full generosity—attributing to him everything that the ancient world recognizes as his.[2]

On this basis, here is a likely summary of the content of Anaximander's *On Nature*:

1. The transformation of one thing into another is regulated by "necessity," which determines how phenomena unfold in time.

2. The multiplicity of things that constitutes nature derives from a single origin or principle, called the *apeiron* (ἄπειρον), the "indefinite" or "infinite."

3. The world came into being when hot and cold separated from the *apeiron*. This separation generated the cosmic order. A ball of flame grew around the air and the Earth "like the bark of a tree." This ball then broke apart and was confined inside the wheels that form the Sun, the Moon, and the stars. The Earth was originally covered in water, which slowly dried up.

4. The Earth is a body of finite dimensions floating in space. It doesn't fall because there is no particular direction toward which it might fall. It is "dominated by no other body."

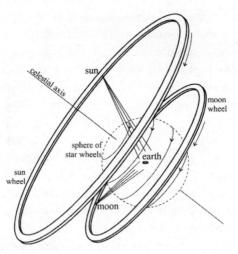

Figure 9. Dirk Couprie's reconstruction of Anaximander's cosmology.

5. The Sun, the Moon, and the stars rotate around the Earth, forming complete circles. Immense wheels, similar to wagon wheels, carry them along (figure 9). They are hollow inside (like a bicycle tire), filled with fire, and pierced along their inward-facing surface. The Sun, the Moon, and the stars that we see in the sky are the fire visible through these holes. The wheels are probably meant to explain why the planets don't fall. The stars are on the wheels closest to us, the Moon on the middle wheel, and the Sun on the wheel farthest from us. Their distances from Earth are in the proportion 9:18:27.*

*Dirk Couprie has hypothesized that these numbers are simply expressions for "very far," "farther still," and "exceedingly far." Others have attempted to justify them as arbitrary measures for the description of a mechanical model, as if we were to say, "Imagine the Moon on a large wheel, and the Sun on a wheel twice as large"—that is, "on a larger wheel."

Figure 10. A hypothetical reconstruction of Anaximander's map of the world.

6. Meteorological phenomena have natural causes. Rainwater is water from the sea and rivers that evaporates because of the Sun's heat. It is carried away by the wind and then falls onto the Earth. Thunder and lightning are caused by colliding and splitting clouds. Earthquakes are caused by fissures in the Earth brought about, for example, by excessive heat or rain.

7. All animals originally came from the sea or from the primal humidity that once covered the Earth. The first animals were thus either fish or fishlike creatures. They moved onto land when the Earth became dry, and they adapted to living there. Human beings, in particular, cannot have been born in their current form, because babies are not self-sufficient, so someone else had to have fed them. They grew out of fishlike creatures.

To these theories we can add the following points:

8. Anaximander drew up the first map of the known world (figure 10). In the generation following him, another Milesian, Hecataeus, expanded this map. Hecataeus's map served then as the basis for all other ancient (and hence modern) maps.

9. Anaximander wrote the first text in prose about natural phenomena. Earlier works on the origin and structure of the world (Hesiod's *Theogony*, for example) had always been written in verse.

10. Anaximander is traditionally credited with introducing the use of the gnomon to the Greek world, perhaps from Babylonia. A gnomon is a rod set vertically in the ground. By measuring the length of the shadow it casts, one can determine the Sun's altitude. A complex astronomy of the Sun's movements can be developed using a gnomon.

11. Some authors report that Anaximander was the first to measure the obliquity of the ecliptic (the path that the Sun appears to trace in the sky during the year). This is possible if, as seems likely, he made systematic use of the gnomon, since the obliquity of the ecliptic is the primary natural measurement shown by the gnomon.[3]

It is hard to reconstruct the general intellectual framework in which these ideas were situated. Gérard Nadaff suggests that Anaximander's primary goal was to reconstruct and explain the history of the present order of things, both natural and social, from the origin of the universe to the present time, using a novel rational and naturalistic perspective. Naddaf notes that cosmological myths have similar aims. Anaximander fits within the tradition of cosmological myths but revolutionizes its method, grounding it in a naturalistic point of view.[4]

No matter what motivated Anaximander's research, one certainly cannot say that the ensemble of these ideas and results constitute a scientific corpus in the sense of modern science. Essential elements of modern science are missing.

For example, the idea of seeking mathematical laws underlying natural phenomena is absent. This concept will appear in the generation after Anaximander, with the Pythagorean school. It will be expanded over subsequent centuries and eventually lead to the grand achievements of Alexandrian science[5]—in particular, to the astronomy of Ptolemy and Hipparchus, a monument of mathematical physics.[6]

Figure 11. An eighteenth-century gnomon, Beijing.

Also, although Anaximander is an acute observer, he lacks the idea of experimentation, in the sense of constructing artificial physical situations for the purpose of making specific observations and measurements. More than two thousand years would pass before the mature form of this idea came to full fruition in the work of Galileo Galilei, to become a keystone of modern science.

The list of differences between Anaximander's theories and modern science could continue: many aspects of Anaximander's vision are decidedly archaic.

Nonetheless, these archaic elements can easily mask the far-reaching conceptual newness of Anaximander's theories and their immense relevance to the development of scientific thinking. In the following chapters, I discuss these contributions, their significance, and their legacy, as they appear not to historians of ancient Greek culture but to a scientist of today.

THREE

ATMOSPHERIC PHENOMENA

Before examining the major theme of Anaximander's cosmology and the subtle theme of the nature of the *apeiron*, I think it is critical to start from an aspect of Anaximander's ideas that is often treated as marginal but seems to me of central importance: his reading of atmospheric phenomena in naturalistic terms.

We know, for example, from Hippolytus that,

[Anaximander holds that] rain derives from the vapor that, under the effect of the Sun, rises from the Earth.

From Aetius and Seneca we learn,

Thunder, flashes, bolts of lightning, hurricanes, and typhoons: According to Anaximander, all of these phenomena derive from the wind.

And what is wind? From Aetius, once again:

According to Anaximander, wind is a current of air, whose subtle and wet parts are set in movement and mixed under the Sun's influence.

Ammianus Marcellinus conveys to us Anaximander's explanation of earthquakes:

> Anaximander holds that the Earth, made arid by an excess of dryness due to great heat, or following the wetness due to abundant rains, is rent by great crevices. An immense quantity of air from above floods into these crevices with violence. The Earth shakes from the violence of the air circulating there. This is why these terrifying phenomena occur during periods of great heat or following overly abundant rain.[1]

And so on. (These texts are typical examples of the form of our sources on the content of Anaximander's book.)

If we read these ideas of Anaximander on natural phenomena in the general context of Greek culture, they appear only to confirm the Greek world's interest in atmospheric phenomena—already expressed, for example, in religion. If we read them in the light of current knowledge and full modern awareness of the physical nature of meteorological phenomena, they appear to us just naïve attempts to explain various phenomena—some mistaken (earthquakes do not occur when it rains too much or when it is too hot), others correct (the origin of rainwater is indeed the evaporation of water from the surface of the Earth).

Both views are short-sighted. They disregard the fact that in all texts earlier than Anaximander that have come down to us, Greek and otherwise, natural phenomena like rain, thunder, earthquakes, and wind are always explained solely in mythical and religious terms: as manifestations of incomprehensible forces attributed to divine beings. Rain comes from Zeus, wind from Eolus. It is Poseidon who stirs up the waves of the sea. Before the sixth century BCE, there was no sign of any attempt to think of these phenomena as tied to natural causes, independent of the will and decisions of gods.

At a certain point in humanity's history, the idea came into being that it was possible to understand these phenomena—their interrelation, causes, and connections—without recourse to the caprices of gods. This immense turning point took place in Greek thought of the sixth century BCE, and it is consistently attributed to Anaximander in all of the ancient texts.

I think that this epochal passage has been overlooked for two reasons.

On the one hand, ancient authors recognized the novelty of Anaximander's naturalistic approach and evoked it frequently and accurately. But the naturalistic explanations for these natural phenomena provided by Anaximander and his followers were not yet very successful in ancient times. Greek science would become outstandingly successful in giving an exact mathematical account of astronomical phenomena like the movement of the Sun, Moon, stars, and planets; would shed light on statics and optics; would establish the bases for scientific medicine, and much more. But it would have limited success in giving solid explanations for complex physical phenomena such as those in meteorology. Therefore, Anaximander's theory of naturalistic explanation remained only a hypothesis for ancient authors, not yet an established explanation of these phenomena.

This is evident in the few texts cited, none of which says, "Anaximander *understood* that," say, rainwater comes from the evaporation of water on Earth. Instead, we read, "Anaximander *holds* that," or "according to Anaximander." In the ancient world, it is not yet clear whether Anaximander's theory for explaining atmospheric phenomena on a naturalistic basis is effective.

On the other hand, today we consider it perfectly obvious that atmospheric phenomena have natural causes, so much so that we overlook the enormous conceptual leap forward that was necessary to formulate this hypothesis.

In Greek religion, the sky was the privileged location of the divine, and meteorological phenomena were the gods' most characteristic mode of expression.[2] Lightning was attributed to Zeus, the father of the gods. Poseidon caused earthquakes. The very unpredictability of meteorological phenomena reflected divine freedom. To seek a naturalistic interpretation for these phenomena, one independent of the gods, represented an enormous break with a religious reading of the world.

In *The Clouds*, written two centuries after Anaximander, Aristophanes shows that Anaximander's naturalistic explanation for thunder and lightning was still perceived as blasphemous toward Zeus:

> STREPSIADES. I always thought that rain came from Zeus. But tell me, who is it makes the thunder, which I so much dread?
>
> SOCRATES. The clouds, when they roll one over the other.
>
> STREPSIADES. But how can that be? You are most daring among men!
>
> SOCRATES. Being full of water, and forced to move along, they are of necessity precipitated in rain, being fully distended with moisture from the regions where they have been floating; hence they bump each other heavily and burst with great noise.
>
> STREPSIADES. But is it not Zeus who forces them to move?
>
> SOCRATES. Not at all. It's the aerial Whirlwind.
>
> STREPSIADES. The Whirlwind! Ah! I did not know that.

So Zeus, it seems, has no existence, and it's the Whirlwind that reigns in his stead?

The comedy ends with the thrashing of Socrates and his friends, corrupters of youth and blasphemers:

STREPSIADES. And with what boldness you insulted the gods, and went seeking as far as the site of the moon? These blows and stones are for you! Come on, let's beat them! They deserve it for so many reasons, above all for their insults to the gods![3]

Aristophanes's comedy is fun, and they say that Socrates (the real one), following the first performance, rose in a friendly spirit to signal his presence and wave at the audience. Plato, too, in the *Symposium*, describes Socrates and Aristophanes dining together affably. Still, twenty-five years later, Socrates was tried by an Athenian court and sentenced to death for corrupting youth with his teachings and failing to acknowledge the city's gods—that is, precisely for the charges that Aristophanes made against him in *The Clouds*. The accusation was that Socrates believed that meteorological phenomena could be understood without recourse to the gods as natural events—precisely Anaximander's temerarious hypothesis.

The opinion that rain could be caused by movements of the wind and the heat of the Sun without the direct intervention of Zeus was probably as disconcerting to a devout Greek of the time as the idea that the human soul is simply the result of the interactions of atoms to a devout Catholic of today. With one difference, however: today's Catholic grapples with a naturalism already twenty-six hundred years old, while Anaximander, as far as we can tell, was the first to propose this kind of naturalistic reading of the world. I will return to the prob-

lems between religion and Anaximander's idea of naturalistic explanations in the final chapters.

Cosmological and Biological Naturalism

Anaximander's naturalistic proposal goes far beyond
meteorological phenomena. To appreciate it fully, we can
compare the description of the origin of the world given
by Hesiod (cited in chapter 1: "Verily at the first Chaos
came to be, but next wide-bosomed Earth . . .") and the
one given by Anaximander, summarized in point 3 of
chapter 2 ("The world came into being when hot and
cold separated from the *apeiron*. . . . A ball of flame grew
around the air and the Earth 'like the bark of a tree.'").
Daniel Graham has recently made a close comparison
between these descriptions, and I summarize his conclusions here. On the one hand, there is an obvious similarity of intent in seeking to describe the origin of the world
and to trace the history of the world. This similarity
shows the continuity of the problem and the cultural
roots of Anaximander's basic interests. On the other
hand, the direction each followed in seeking a solution
could not be more different. Hesiod, as I underlined in
chapter 1, fits easily within the universal tradition: he
tells the history of the world as a series of stories about
the gods and their personal relationships. Anaximander
breaks brutally and radically with this tradition. In his
history of the world, there is no trace of supernatural.
The things of the world are explained in terms of worldly things: fire, cold, heat, air, earth. The matters requir-
ing explanation are worldly things: the Sun, the stars, the
Earth.[4]

An inattentive reader could even easily mistake
Anaximander's history of the world for an approximate

version of the history of the Big Bang as told by modern cosmology. It is important not to mistake this vague similarity for some mysterious foresight on the part of Anaximander: nothing of the sort. The similarity is neither accidental nor mysterious: Anaximander makes a precise methodological proposal to explain the cosmos—explaining the facts of the world in terms of the things of the world—and to this day, we continue to develop his methodological proposal, which has proven to be effective. The story of the Big Bang, like Anaximander's story, and unlike all other cosmologies of all civilizations, is an attempt to understand the history of the world solely in terms of natural things, without reference to the divine. (There is another area in which Anaximander's naturalism appears to be of nearly miraculous success: his reflections on the origins of life and human beings. According to Anaximander, life begins in the sea. He refers unequivocally to an evolution of living species connected with the evolution of climate conditions. First come marine species, which, as the earth dries up, migrate and adapt to dry land. He wonders which living beings might have brought forth the first humans. These issues would return only in recent centuries, with Darwin, with the momentous results we all know. With all their limits, the fact that these ideas are already present in the sixth century before the Common Era is breathtaking.)

Even if the actual explanations proposed by Anaximander were mistaken, the very fact of his proposing research into natural causes and explanations for atmospheric phenomena marks the birth of scientific inquiry in the world. But Anaximander's explanations are not all mistaken. On the contrary, most are surprisingly accurate. The origin of rainwater is indeed the evaporation of water on Earth caused by the heat of the sun. The key physical event in an earthquake is indeed the fracturing of the earth. Life did indeed begin in the seas and evolve to exist on land. How did Anaximander manage to understand all this?

The key is perhaps simple skepticism vis-à-vis the usual explanations. A century after Anaximander, Hecataeus of Miletus, who built on Anaximander's map and was the first Greek historian, opens his *Genealogiai* with a famous incipit: "So speaks Hecataeus of Miletus: I write that which I consider true; for the many stories that the Greeks tell are in contradiction among themselves and seem ridiculous to me."[5]

Once the idea of seeking naturalistic explanations has been formulated and once this healthy skepticism has been kindled, a certain number of reasonable explanations follow directly from the simple observation of the world.

But let us recall how, in elementary school, we all learned the cycle of water, with wonder, from our schoolbooks. A drop of water falls as rain, flows into rivers, reaches the sea, evaporates under the sun's heat, is carried off by wind, and falls again as rain. It is a stupendous example of the complexity but, above all, the comprehensibility of our beautiful world. Schoolbooks do not say, but the one who understood the cyclical route of water was Anaximander of Miletus.

FOUR

EARTH FLOATS IN SPACE, SUSPENDED IN THE VOID

Many people today believe that in medieval Europe, the Earth was thought to be flat. According to this legend, when Christopher Columbus proposed to travel to China by sailing west, he was opposed by Spanish scholars who deemed his undertaking absurd because, in their view, he would fall off the edge of the Earth.

This legend is without substance. It is odd that it has endured in my country, Italy, where every schoolchild studies *The Divine Comedy*, a summa of medieval knowledge written two centuries before Columbus. In *The Divine Comedy*, Dante describes with great visual clarity an obviously spherical Earth. No one in medieval Europe believed the Earth to be flat. Saint Augustine, for example, argued that the existence of men living at the antipodes was impossible for reasons having to do with their relationship to Jesus Christ, but he did not challenge the idea that the Earth was spherical. At the very beginning of the *Summa Theologica*, Saint Thomas

Aquinas refers clearly to the Earth's spherical form.[1] There are almost no medieval texts that refer to a flat earth.*

In contrast, the objections raised by the scholars at the Spanish court to Columbus's plans were anything but unfounded. In the year 1400, the precise size of the Earth was known, with a margin of error of a few percent. It had been known since the third century BCE, when Eratosthenes, the director of the Great Library of Alexandria, measured it using a brilliant theoretical and observational technique. The Earth was too big to be circumnavigated without stopovers using the naval technology available in Columbus's day. Columbus tried to convince the Spanish court that the Earth was smaller than it really is, and that it was therefore possible to sail to China taking a western route without depending on known ports for food and water. In simple words, Columbus was wrong. Columbus died believing that the Earth was small and he had arrived in Asia. Of course, the twists of fate are unforeseeable, and Columbus's error determined the course of history (including, for instance, the extermination by Europeans of some 20 percent of humanity over the course of the following decades).

The belief that the Earth is a sphere was already established in Greece in Aristotle's time. Aristotle's writings

---

*There are rare exceptions: Lactantius in the fourth century CE, and Cosmas Indicopleustes in the sixth century CE. They were Christian authors who, in their zeal to categorically reject pagan thought, sought without success to return to the archaic idea of a flat Earth. According to Cosmas, the Earth is shaped like the Ark of the Covenant.

on the subject and the arguments he makes in support of the Earth's spherical form are correct and convincing to any person of good sense who takes the trouble to read and think through them. Should any doubts remain, the lucid first chapter of Ptolemy's *Almagest* offers complete and definitive clarity on the subject. Since shortly after Aristotle's time, no one in the West challenged the fact that the Earth is (more or less) spherical.

A generation earlier than Aristotle, the concept of a round Earth was already well known, but there was less clarity on the issue. Plato in the *Phaedo* has Socrates say that he maintains that the Earth is a sphere,[2] but he adds, "I myself should never be able to prove it." This passage in the *Phaedo* is the oldest direct evidence we have of the belief in a spherical Earth.

The conceptual clarity on this exquisitely scientific topic in the Greece of the fifth century BCE is impressive. Plato and Aristotle make a neat distinction between maintaining a point and possessing convincing scientific arguments in support of it. I think that the average educated European or American of today knows that the Earth is round, but is probably not able to offer direct and convincing proof of this belief. His level of scientific understanding, at least as far as this topic is concerned, lies somewhere between Plato's generation and Aristotle's.

There is another consideration that is perhaps interesting in this regard. The *Phaedo* is one of the most read, taught, and discussed texts in philosophy. But almost anyone who comments on it focuses solely on the soul's immortality and fails to notice that it contains this jewel of the history of science: the first written evidence we have of the new worldview, with a spherical Earth. This is glaring evidence of the abyss between the sciences and humanities in our time, each stupidly blind to the other.

Plato mentions that the Earth is round as if it was already a well-known idea. Where did the idea come from? It is sometimes attributed to Parmenides, but more often considered of Pythagorean origin, possibly going back to Pythagoras himself. Anaximander did not imagine the Earth to be round; instead, he refers to a more or less cylindrical shape, like that of a shallow drum or thick disk: "[Anaximander says that] the Earth is suspended in the void, supported by nothing, but stable because of its equal distance from everything. Its shape is rounded, like a column in stone. It has two surfaces, one made of the ground beneath us, and another opposite this."[3]

This cylindrical, disklike form may seem strange. I believe that one plausible explanation for it is as follows. Thales had taught that water was the origin of all things and imagined an immense ocean from which everything is born and upon which the Earth floats. Thales's Earth is a floating disk; its round shape follows the ancient idea that the emerged land forms a circle surrounded by sea. Anaximander's insight is that the ocean supporting the Earth isn't needed. Without the ocean, he is left with a disk floating in space.

Now, the point generally overlooked but of main importance for understanding Anaximander's achievement is the following. From a scientific perspective, the key step forward is not establishing whether the Earth is cylindrical or spherical; it is understanding that the Earth is a finite body that floats free in space. I examine this point in detail, because its significance can easily escape those who do not have direct experience in scientific research.

The Earth, in reality, is neither a cylinder nor a sphere. It is an ellipsoid slightly flattened at the poles. In truth, it isn't even an ellipsoid but rather a kind of pear, since the South Pole is more flattened than the North Pole. In fact, it is not pear-shaped either, because today we can detect additional irregularities. These progressive refinements in our understanding of the Earth's exact form are of interest to some, but in and of themselves they add nothing essential to our understanding of the world. The passage from Anaximander's cylinder to the sphere, ellipsoid, pear, and finally irregular form of today represents a progressive refinement in our knowledge of our planet's form, but it is not a conceptual revolution.

By contrast, understanding that the Earth is a stone that floats unsupported in space, with the same heaven underneath it as the one we see above—*this* is a huge step forward conceptually. And this is Anaximander's contribution.

Anaximander's cosmological model, with a cylindrical Earth, is often presented by scholars who lack a developed scientific training as primitive and uninteresting,[4] while the Pythagorean/Aristotelian model, with the spherical Earth, is presented as "scientifically correct." Both of these judgments reflect scientific illiteracy, for opposite reasons. First, as noted, the conceptual leap from a flat Earth to a finite Earth floating in space is immense and arduous. The fact that the Chinese Imperial Institute of Astronomy in two thousand years of existence failed to make this leap proves its difficulty. No other civilization made it either. By contrast, the conceptual leap from a cylindrical Earth to a round Earth is easy. The proof? It happened in only one generation. Second, as noted, the spherical model is by no means the "true" answer to the question of the Earth's shape. It is somewhat more precise than the cylindrical model and somewhat less precise than the ellipsoid model.

To Anaximander, then, without the slightest doubt, goes the full merit of the first great cosmological revolution.

But how was Anaximander able to understand that there is sky beneath the Earth?

A moment of reflection shows that there is plenty of evidence for this idea. Every evening, the Sun sets in the west; the next morning, it reappears in the east. How does it go from west to east during the night? Consider the North Star. On a clear summer night, we see all the other stars revolving slowly and majestically in the sky, while the North Star stays still, as a pivot. The stars closest to the North Star—the stars of the Little Bear, for example—revolve around it slowly and complete the circle in (approximately) twenty-four hours. They are always visible in the heavens (when we are not blinded by sunlight, that is). The stars a bit farther from the North Star complete a larger orbit in twenty-four hours, and the size of the orbit increases along with their distance from the North Star, until they seem to brush against the horizon to the north.

At times, a star seems to disappear behind a mountain and reappear slightly to the east a short time later (figure 12). Manifestly, it passes behind the mountain. And those farther from the North Star? They, too, seem to disappear behind something and then reappear. For them to be able to trace this route, there must be space *down there*. What of the stars on the celestial equator, far from the North Star, that are near the Sun's path in the sky? Doesn't one think immediately that they, too, disappear

Figure 12. An extremely long-exposure photo of the night sky show-ing the movement of the stars over the course of the night around the North Star. The photo shows clearly that, beneath the horizon, there must be empty space in which the stars can complete their orbit.

behind the Earth and pass beneath it? And if they pass beneath the Earth, there must be empty space beneath the Earth!

Notice how the structure of this discovery resembles the discovery that rainwater comes from evaporation. In one case, water disappears from a bowl left out in the sunshine and appears falling from the sky. Intelligence connects disappearance with reappearance and identifies rainwater with evaporated water. In the other case, the Sun disappears in the west and reappears in the east; intelligence connects disappearance with reappearance and seeks the route connecting them: the empty space beneath the Earth. It is nothing more than the combina-tion of curiosity with clarity of intelligence.

In grasping that there is a void beneath the Earth, indeed, Anaximander uses nothing more than the simple inference we make when we see a man disappear behind a house and reappear on the other side. How could that happen? There must be an open passage behind the house. Easy.

Easy? If it were really that easy, then why did generation after generation of human beings not come to the same conclusion? Why did so many civilizations go on believing that beneath the Earth there had to be more earth? Why did the Chinese, despite the splendor of their ancient civilization, not grasp this fact until the Jesuits arrived in the seventeenth century? Was the world outside of Miletus full of idiots? Certainly not. Why, then, was this point so difficult to grasp?

The difficulty derives from the fact that the idea that the Earth floats in space contradicts our fundamental experience of the world. In light of our experience, the notion is obviously absurd—unheard of—unbelievable.

First, we must accept that the world may not conform to our direct experience and to our long-held image of it, that things may be other than they seem and from the way everybody has always thought they are. We must let go of an image of the world that is familiar to us. What is needed to take this step is a civilization in which human beings are ready to call into question what everyone has always believed to be true.

Second, we must construct a credible and consistent alternative to the old image of the world. The fact that the Earth floats contradicts the rules that we know regulate the world: objects fall. If nothing were holding it up, the Earth would fall. If the Earth isn't supported by anything, why does it not fall?

Marking deductions from available evidence and supposing that there is nothing beneath the Earth was not the hard part. This idea may have come up in the history of Chinese astronomy and, very possibly, elsewhere. But in science it is not difficult to come up with ideas; it is difficult to come up with workable ideas, to find a way to compose and articulate new ideas as part of a whole that is consistent with the rest of our knowledge, and to convince others that the entire process is reasonable. What is difficult is to have the courage and intelligence to conceive and articulate a new, coherent, overarching image of the world.* It was difficult to reconcile the idea of the Earth suspended in the sky, which accounts neatly for the daily movement of the stars, with the obvious, experiential fact that any heavy object falls.

The genius of Anaximander is that he takes on the question, Why, then, does the Earth not fall? Aristotle relays his answer in *De Caelo* (*On the Heavens*). In my

---

*Like many scientists, I have drawers and files filled with mail from people who write to me with new scientific ideas, original and daring, but useless. Ideas can come and go many times, but an idea on its own is useless. In the third century BCE, Aristarchus considered the possibility that the Earth rotated on its own axis and around the Sun. In light of the Copernican revolution, his idea was correct. Still, Copernicus and not Aristarchus deserves the credit for this revolution, because it was Copernicus who showed how this idea might work and how it could be integrated with the rest of our knowledge; he set in motion the process that eventually persuaded the rest of the world. It is easy to have ideas; it is difficult to pick out good ideas and find the arguments to show that they are "better" than current notions. Who knows how many human beings had imagined that the Sun passed beneath the Earth without, however, being able to change humanity's worldview.

opinion, this is one of the most beautiful moments in the history of scientific thinking: the Earth does not fall because there is no particular reason for it to fall. In the words of Aristotle:

> There are some, Anaximander, for instance, among the ancients, who say that the earth keeps its place because of its indifference. Motion upward and downward and sideways were all, they thought, equally inappropriate to that which is set at the centre and indifferently related to every extreme point; and to move in contrary directions at the same time was impossible: so it must needs remain still. This view is ingenious.[5]

The argument is extraordinary and perfectly correct. Aristotle sees this: he does not credit many to be "ingenious." What, precisely, is the argument? It consists in overturning the question, Why doesn't the Earth fall?, transforming it into, "Why should the Earth fall?" The genius of Anaximander, in modern words, is to question the extrapolation from the objects of our experience to the Earth itself, of the observed universality of falling. More precisely, to take the observational evidence from the motion of the Heavens as an argument against the legitimacy of this extrapolation. This is science at its best. The point is even clearer if we read Hippolytus, whom we can translate as, "The Earth is aloft, not dominated by anything; it remains in place because of the similar distance from all points."[6]

In our everyday life, heavy objects fall, but they are in the vicinity of an immense body—the Earth—that "dominates" them and determines a preferred direction: toward the Earth. But Earth does not have any particular direction in which to fall because nothing "dominates" it. Objects do not fall in the direction of an absolute "down," a single direction that is the same

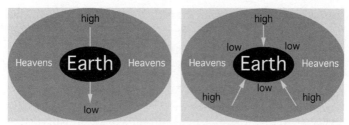

Figure 13: Anaximander's basic insight: the universe does not resemble the image on the left, and there is no privileged direction (here called "high-low") that determines how things fall. The figure on the right is a hypothetical illustration of the idea that an object's fall is determined by the presence of something that "dominates" it (the Earth), which determines a privileged direction (toward the Earth). We do not know if Anaximander could have drawn a figure like this, and the shape of the Earth in these drawings does not necessarily reflect that imagined by Anaximander.

throughout the universe; objects fall toward the Earth if they are on the Earth's surface.[7]

Notice, then, that the very meanings of "up" and "down" become ambiguous. We can continue to say that objects fall "down," however "down" no longer indicates an absolute direction in the cosmos (see figure 13).[8] Another text by Hippolytus is explicit on the matter: "For those standing on their own two feet down below (at the antipodes), high things are low, while low things are high . . . and so it is all over the Earth."[9]

The concepts of "high" and "low," or "up" and "down," structure our direct experience of the world and form the basis of our mental organization of the physical world. In the new world posited by Anaximander, the meaning of these concepts changes in depth. In bringing about his revolution, Anaximander has to understand that the notions of "up" and "down" required to make sense of the universe and to determine the direction of falling differ from those of our everyday experience. "Up" and "down," as commonly understood, do not constitute an absolute and universal structure of the physical world or a preexisting structure of space. "Up" and "down" are not absolute: they do not apply to the Earth itself.*

The deep change in perspective on the cosmos engineered by Anaximander has much in common with other great scientific revolutions. The step forward he took is similar to the one Galileo took that led to the triumph of the Copernican revolution. Does the Earth move? How could it move, when it seems evident that it stands still? No—Galileo understands, completing the Copernican revolution—absolute motion and stasis do not exist. Objects resting on the Earth are immobile with respect to one another, but this does not mean that, as a group, they cannot be in motion within the solar system. The concepts of "stasis" and "motion" are much more complex than our everyday experience indicates. Similarly, with his theory of special relativity, Einstein understands that the idea of simultaneity—of "now"—is not absolute either, but instead relative to the observer's state of motion.

The difficulty in understanding the complexity of the notion of simultaneity in Einstein's theory is very much

*This does not imply that Anaximander believed that the Earth is the cause of the fall (as in Newton), nor that the position of the Earth is caused by the radial direction of the fall of heavy objects (as in Aristotle). Anaximander, like Copernicus, might have had no dynamical theory of falling at all.[10] Aristotle, indeed, criticizes Anaximander precisely for this: for failing to see what Aristotle considers his own great insight into the problem, which is not the older idea that things fall toward the Earth, but the beautiful idea that the Earth is at the center of the universe because of the natural tendency of heavy objects to fall toward the center—a natural tendency introduced by Aristotle.

analogous to the difficulty in understanding the notions of "up" and "down" in Anaximander's new cosmological theory. If the relativity of "up" and "down" nowadays seems fairly easy to understand, while the relativity of simultaneity is still harder to grasp for those who are not professionals physicists, this is only because Anaximander's theory and its developments have been digested for twenty-six hundred years, while Einstein's is not yet widely assimilated. But we are dealing with the same conceptual path. The difference is that Einstein based his work on observations already fully codified in Maxwell's theories and the mechanics of Galileo and Newton, while Anaximander based his only on the observation of the rising and setting of the stars.

Anaximander's greatness lies in the fact that on the basis of so little, in order to better account for his observations, he redesigns the universe. He changes the very grammar of our understanding of the universe. He modifies the very structure of our conception of space.

For centuries, human beings had understood space as intrinsically structured in the direction toward which objects fell. No, says Anaximander: the world is not as it seems to us. The world is different from how it appears. Our perspective on the world is limited by the smallness of our experience. Reason and observation allow us to understand that our prejudices about the world's functioning are mistaken. Space does not have a privileged direction toward which objects fall. For the Earth itself, there is no "down" toward which it might fall.

This is a dizzying conceptual tour de force—and it is correct. Once a coherent conception of the world has been formulated in which objects fall not toward an absolute "down" but toward the Earth, there is no longer any reason for the Earth itself to fall. The focal point of Anaximander's argument, conveyed by the texts that have come down to us, is that the expectation that the Earth must fall is based on an unjustified extrapolation.*

Intelligence, used well and in conjunction with observation, frees us from an illusion, from a limited and partial view of the world. It remakes our understanding of the world in a new form. This form is more effective. To be sure, it can be improved: going forward, humans will have to learn that the Earth is not a drum but a sphere; that it is not really a sphere; that it is not at rest but in perpetual motion; that the Earth attracts other bodies; that all bodies, in fact, attract each other; that this attraction is the very curvature of space-time, etc. Each of these steps will take centuries, but the process has begun. It has been set in motion by a first great step, one that overturns a conception of the world common to all civilizations and brings forth the conception of a spherical world, surrounded by the sky, the distinctive mark of Greek civilization and of all civilizations, like our own, who are heirs to the Greeks.

There is another important novelty in Anaximander's cosmology, emphasized by Dirk Couprie. The heavenly vault had always been seen as the upper enclosure of the world. Humanity had seen Sun, Moon, and stars as enti-

---

*This exquisitely scientific argument can be hard for philosophers and historians to grasp. One reads, for example, that "we must wait for Newton to have the correct answer to the question of why the Earth does not fall." This is utter silliness. Why is Newton's answer the right one? Simply because it is the one that we learned at school, given that Kepler was no longer in fashion and Einstein was not yet taught at our school? There is no sense in which the problem of why the Earth does not fall was solved better by Newton than by Anaximander, Aristotle, Copernicus, or Einstein. Each one of these names represents a step toward a more powerful conceptualization of the world.

ties that traveled along the celestial vault itself—the ceiling of the world—all at the same distance from us. Anaximander, observing the heavens, for the first time did not see a vault, but instead an open space in which heavenly bodies were at various distances from us. The numbers that he proposes for the spokes of the wheels of Sun, Moon, and stars matter less for their specific values than for suggesting the possibility that these numbers may mean something. The step is from the world seen as the inside of a box to the world immersed in an open, external space.* As Couprie says, Anaximander in some way invents the open space of the cosmos.[11] The ramifications of this conceptual innovation are immense.

In the history of science, perhaps the only other example of a conceptual revolution comparable in greatness to Anaximander's is the Copernican revolution, opened by the publication of Copernicus's treatise in 1543.† Like Anaximander, Copernicus rethinks the map of the cosmos. In place of a cosmos made up of the Heavens above and the Earth below, Anaximander puts forth an open cosmos where the Earth floats, surrounded by the heav-

*Couprie asked me whether I, as a physicist, can understand the logic that led Anaximander to infer that the Sun, Moon, and stars were located at different distances from Earth. The only answer I can come up with is that if they were the same distance away, the wheels that carry the various celestial bodies (required to account for the fact that they don't fall) would pass one through the other, which does not make sense.

†The title of Copernicus's book is *De revolutionibus orbium cœlestium*, or "On the Revolutions of the Celestial Bodies," where "revolution" means the circular motion of the planets in the sky. Because Copernicus's book embodies the greatest scientific upheaval, it is hard to resist the idea that the word "revolution" acquired its meaning of "major upheaval" under the influence of the book itself, even if, according to some etymological dictionaries, the use of the word in the sense of "instance of great change in affairs" is recorded from the mid-fifteenth century, a bit before Copernicus's book.

ens. Copernicus moves this floating Earth from the center of the cosmos to an orbit around the Sun. As was the case with Anaximander, the Copernican revolution paved the way for immense scientific developments that would occur over the course of the following few centuries.

There were other similarities. Copernicus studied in Italy—a land of political disunity, trade, and openness to the rest of the world. He was nourished by the rich, vibrant cultural ferment of the early Renaissance. Anaximander emerged from the new cultural climate of young Greek civilization, similar in many respects to the Italian Renaissance.

But again, Copernicus based his theories on the vast technical and conceptual work accomplished by the Alexandrian and Arab astronomers. Anaximander's work was based on nothing more than the first questions and the first imprecise speculations of Thales, and on what he had observed with his own eyes—nothing more. On this slight basis, Anaximander achieved what I think must be deemed the first and greatest of all scientific revolutions: the discovery that the Earth floats in an open space.

I close this chapter with two quotations; the first is from Charles Kahn: "Even if we knew nothing else concerning its author, [Anaximander's theory on the Earth's position] alone would guarantee him a place among the creators of a rational science of the natural world."[12] The second is from Karl Popper, among the greatest philosophers of science of the twentieth century: "In my opinion this idea of Anaximander's [that the Earth is suspended in space] is one of the boldest, most revolutionary, and most portentous ideas in the whole history of human thinking."[13]

FIVE

INVISIBLE ENTITIES AND NATURAL LAWS

Philosophy school textbooks tell us that the first philosophical school was the Ionic school, which consisted of Thales, Anaximander, and Anaximenes. These philosophers sought the "first principle" as the basis of all things, which Thales identifies as water, Anaximander as the *apeiron*, and Anaximenes as air. Presented in this manner these ideas are incomprehensible, and leave one wondering how such three bits of nonsense could have given birth to philosophy. I add here some details, trying to make these philosophers' thoughts about the first principle more reasonable, but from a scientific rather than philosophical viewpoint.

Before considering Anaximander, it is useful to look at Thales (in the generation before Anaximander) and Anaximenes (probably in the generation after Anaximander).

THALES: WATER

We know little of Thales as well. He is said to have traveled widely and to have participated in civic life in Miletus. Some theorems in elementary geometry along with their proofs are attributed to him. His most important conceptual achievement is held to be initiating the search for the *arche* (ἀρχή), or "origin," a principle that can explain all natural phenomena. The precise meaning of *arche* in the Milesian school has inspired much debate. I do not have the competence to address these issues here; I wish simply to reflect on the relevance of this question to the subsequent development of science.

I think it is useful to look for the meaning of an expression in the way it is used. In and of itself, "principle" may mean all sorts of things, and focusing on the a priori meaning of a word is not particularly clarifying. The meaning of *arche* does not become clearer in trying to understand the metaphysical concerns that drove Thales to search for a principle: it does clarify, instead, when we observe what Thales *does* with this concept.

What he does with the concept of "principle" is simple. He tries to bring together the immense variety of natural phenomena that can be observed by relating them within a unifying explanation, intrinsic to nature itself. He attempts to find a simple explanation for how nature works. Considered in this light, the process delineated by Thales is nothing else than the process of scientific thinking. The coarseness and naïveté of the specific explanation he considers ("everything is made of water") reflect only early difficulties and the rudimentary state of a first attempt to realize such a program.

Thales probably inherited the idea of the fundamental importance of water and the sea from the mythological

world. As noted earlier, he sees the world as a disk floating on water. This image is probably Mesopotamian in
origin and may be related to the idea, widespread
throughout the ancient world, that in whatever direction
one travels, one arrives at the sea (the river Oceanus that
surrounds all emerged lands). In the *Enuma Elish*, cited
earlier, the universe arises from the liquid chaos of the
waters of the god Apsu.

The opening lines of Genesis, in the literal translation
by E. A. Speiser, read,

> When God set about to create heaven and earth—the
> world being then a formless waste, with darkness over the
> seas and only an awesome wind sweeping over the
> water—God said, 'Let there be light.' And there was
> light.[1]

Notice that before the creation of light, the seas were
already there. The *Iliad*, too, calls Oceanus the father of
the gods. The idea may be even older and may have originated prior to the separation of Eurasian and American
peoples. Consider the first verse of the Navajo creation
myth: "The One is called 'Water Everywhere.'"[2]

Thales could therefore have taken the idea that all
things have their origin in water directly from mythology, or from his travels to Babylonia. But the way he inter-
prets the role of water is distinctively nonmythical and
nonreligious. Thales's water is normal water. Thales's
ocean is not a god.

Thales's naïve attempts to explain natural phenomena
illustrate the methodological and naturalistic clarity of
these earliest glimmerings of rational thought—and
their distance from mythology. For example, Thales suggests that earthquakes are due to the movements of the
Earth, shaken by waves as it floats on the water.

Thales's work is riddled with problems (How is it that the Earth floats, and does not sink?), but it contains the seeds from which Anaximander's splendid, naturalistic explanations will grow.

ANAXIMENES: COMPRESSING AND RAREFYING

Anaximenes's achievement in replacing Thales's water (and Anaximander's *apeiron*, to be discussed shortly) with air lay not so much in the choice of the element air but in his successful attempt to grapple with an obvious problem in the theories of Thales and Anaximander. If everything is made up of water or *apeiron*, how is it possible that water or *apeiron* take on so many different forms and consistencies, as seen in the immense variety of natural substances? How can a primal substance assume such diverse characteristics? Later, the problem would be pointed out by Aristotle. Using the typical language of Greek physics, Aristotle asks how the same substance could appear as light sometimes and heavy at other times.[3]

Simplicius tells us Anaximander's unimpressive attempt at solving this problem: "[According to Anaximander], things do not originate from a change in the underlying substance, but by separation; for the opposites existing in the substance are separated because of the eternal movement."[4]

By "opposites," Simplicius means hot and cold, wet and dry, etc. Not a very convincing answer.

Anaximenes seeks a more reasonable mechanism that could allow a single substance to assume different appearances. With admirable acumen, he identifies this mechanism in compression and rarefaction. He hypothesizes that water is generated by compressing air; air, in turn, is obtained by rarefying water. Earth is generated by further compressing water, and so on for other substances. This is definitely a good step forward toward a more rational understanding of the world's structure.

Later Ionic thinkers added to Anaximenes's idea of compression and rarefaction the idea of a small number of primary substances, combinations of which can give rise to many different materials. Leucippus and Democritus, the atomists, made the notions of compression and rarefaction more concrete and understandable by introducing the idea of elementary atoms that move about in the void.

Notice that today we believe that essentially all of the matter that we normally encounter is made up of three components: electrons, protons, and neutrons. The diversity of the matter forms we perceive is entirely determined by the different combinations and the greater or lesser rarefaction or compression of these few components.

Once again, though, reading these similarities between Greek science and modern science as mysterious prescience on the part of Greek thinkers would be missing the point. The fact is that some general theoretical schemes elaborated in the first centuries of Greek civilization have proven effective. Today, centuries later, they continue to function very well. As always, the challenge lies not in finding answers, but in asking the right questions. Asking if there is some simple stuff making up all matter, and what can be the mechanism generating variety—this was a very good question.

ANAXIMANDER: APEIRON

Let us return to Anaximander, who came before the conceptual evolution introduced by Anaximenes. What is this *apeiron* that, in Anaximander's view, makes up the world?

The issue has inspired much debate, and opinions oscillate between two extremes established by the meanings of the Greek work *apeiron*: the first, "without confines" or "infinite"; and the second, "without fixedness," "indistinct," "undifferentiated."

Once again, I do not wish to enter the discussion on the precise meaning of the term, because, from a scientific perspective, I find it irrelevant. It is akin to asking whether George Johnstone Stoney, when introducing the term "electron" in 1894, meant a "grain of electricity" or a "new particle," or something else. It matters not a whit why he chose the word "electron"; what matters is the introduction of a new idea, the role that this idea assumed within the theoretical scheme set forth by Stoney and his successors, and its effectiveness in describing the world. If Stoney had given this new entity a different name, history would have unfolded no differently. In fact, in contemporary physics, the closest relatives to electrons are called "quarks," a term introduced by Murray Gell-Mann. "Quark" is a seagull's cry; Gell-Mann chose this name to make it known that he was a cultivated man and had read *Finnegan's Wake*, by James Joyce, where the term is used in the line, "Three quarks for Muster Mark!" The only relationship between the word and the particle is that there are three kinds of quark particles.

In the same fashion, had Anaximander called his principle something other than "infinite" or "indistinct," the scientific relevance of his idea would have been strictly the same.

What is thus the meaning of the idea that Anaximander put forward introducing the notion of the *apeiron*?

The essential feature of the *apeiron* is that it is not one of the substances of our common experience. Simplicius writes, "Anaximander said that the origin of beings is the *apeiron*" and comments:

Anaximander . . . was the first to introduce this name "origin" [*arche*];* he says that it is neither water nor any other of the so-called elements, but some other undifferentiated nature, from which come all the heavens and the worlds that exist in them . . . and he expressed this with rather poetic words. It is clear that, having observed the change of the four elements [water, air, earth, and fire] into one another, he did not think fit to make any one of these the material substratum, but rather something else besides these.[5]

Anaximander thus posits that all substances in our experience can be understood in terms of something that is natural but, at the same time, is not one of the substances in our everyday life. The fundamental insight here is that in order to explain the world's complexity, it is useful to postulate or imagine the existence of something that is not part of the world of our direct experience but that can function as a unifying natural element to explain all things.

On the one side, then, Milesian speculation as practiced by Thales, Anaximander, and Anaximenes liberates nature from being identified as a manifestation of a divine, supernatural reality. We can say that the very notion of "nature" as a field of inquiry is the fundamen-

*To be clear: Thales was the first to posit that "all is made by water," thus introducing the idea that there could be a single underlying substance, or principle, to all things; Anaximander, according to Simplicius, is the first to use the word *arche* to denote such a substance or principle.

tal contribution of the Milesian school. The term used for this meaning of "nature," φύσις (*physis*), probably originated in Miletus. On the other side, the very idea of studying nature is based on the recognition that nature does not reveal itself in its entirety to direct experience. On the contrary, we must probe and investigate its origins and structure. Truth is accessible and an integral part of nature itself, but truth is hidden. The instruments with which it can be reached are observation and intelligence. Thought must be ready to imagine the existence of more natural entities than what we directly perceive.

This is precisely the route taken by theoretical science in the following centuries, until today. In positing the existence of the *apeiron*, Anaximander paves the way for something that science will do again and again with immense success: imagine the existence of entities that are not directly visible or perceptible but that allow us to account for natural phenomena.

Atoms—those of the Greek atomists Democritus and Leucippus, as well as their nineteenth-century relatives, those of John Dalton—are the direct descendants of Anaximander's *apeiron*. They are natural objects (nothing is particularly divine about atoms) that escape our direct perception but in terms of which we understand the constitution of matter.

Michael Faraday's great contribution to modern science is another example. In the midnineteenth century, there was no unified understanding of electrical and magnetic phenomena. Following an in-depth experimental investigation, Faraday conceived the idea of a new entity: the field.

A field is an entity assumed to fill out space, like an immense spider web that reaches everywhere, woven by imperceptible lines, now known as Faraday lines.

Faraday introduces two fields, the electric and magnetic fields (others will follow). These influence each other and are responsible for the electrical and magnetic forces. In an extraordinary passage from his beautiful book, Faraday wonders whether these fields, which permeate physical space, are real themselves. He hesitates, but finally concludes that they are.[6] Newton's vision of the universe—the void of space crisscrossed by particles exerting forces upon each other at a distance—is overturned. A new entity, the field, takes its place in the world.

Within a few years, James Clark Maxwell transforms Faraday's visionary insight into a system of equations describing the fields. He grasps that light is nothing but a swift ripple upon these webs, and that these ripples, at greater wavelengths, can bear signals. Hertz reproduces them in the laboratory, and Marconi builds the first radio. Modern telecommunications are grounded on this new image of the world, where unobservable fields are the key ingredient.

Atoms, the electrical and magnetic fields of Faraday and Maxwell, Einstein's curved space-time, the phlogiston of heat theory, Aristotle's ether as well as Lorentz's, Gell-Mann's quarks and Feynman's virtual particles, the wave function of Schrödinger's quantum mechanics, and the quantum fields that form the foundation of contemporary fundamental physics' description of the world—all these are "theoretical entities" that cannot be perceived directly by the senses but are postulated by science to account for the complexity of phenomena in a coherent way. They have precisely the same role and same function as the ones assigned to the *apeiron* by Anaximander. They are the descendants of Anaximander's vision.[7]

The theory of the *apeiron* is rudimentary and certainly cannot be compared to the detailed mathematical theory that Maxwell writes for the electric and magnetic fields or that Feynman writes for quantum field theory. But when our television is on the blink and the antenna repairman tells us that electromagnetic waves aren't being captured well because of a hill, he is using those waves as "theoretical entities" to account for phenomena. He is using a conceptual reasoning that has a precise historical origin, Anaximander's *apeiron*.

THE IDEA OF NATURAL LAW: ANAXIMANDER, PYTHAGORAS, AND PLATO

Let us look again at the only text by Anaximander that has come down to us, as relayed by Simplicius:

> All things originate from one another, and vanish into one another
>
> According to necessity;
>
> They give to each other justice and recompense for injustice
>
> In conformity with the order of Time.

Explicit in these few lines is the idea that the world's becoming does not unfold haphazardly but is ruled by necessity—that is, by some form of law. Also explicit is the notion that the way these laws are expressed is "in conformity with the order of Time." A clear idea here is that natural laws exist, and that these laws govern how things evolve over time.

The form of these laws is not specified. There is a vague analogy with justice or moral law. As far as we know, none of these laws was explicitly set forth by Anaximander.

It will be during the next generation that another major figure in the history of science understands the form that these laws need to have—or, rather, the language in which they need to be written: Pythagoras. His insight, entirely new with respect to the Milesian school, is that the laws of the universe are written in the language of mathematics. This adds a key new element to Anaximander's worldview and gives a precise form to the notion of "law," still vague in Anaximander.

According to traditional accounts, Pythagoras was born in Samos, near Miletus, in 569 BCE. He was therefore twenty-four when Anaximander died in 545. Iamblichus Chalcidensis, a Neoplatonic philosopher of the third century CE, writes one of the most detailed ancient sources on the philosopher, the *Life of Pythagoras*.[8] In it he records that Pythagoras went to Miletus when he was eighteen or twenty to meet Thales and Anaximander. Iamblichus may not be completely trustworthy, but in the small world of the Greek aristocracy, a tiny universe in which everyone seemed to know everyone else, it is hard to believe that Pythagoras and Anaximander did not meet. They were two men with the same powerful hunger for knowledge who lived at the same time and in the same region. It seems highly improbable that the young Pythagoras failed to take an interest in the ideas of his illustrious neighbor before setting off on the long travels that would lead him to Crotone, in Italy, where he founded his school. The similarities in their cosmological interests and, above all, the idea of Earth afloat in space common to both men make it nearly certain that Pythagorean thinking was influenced by the Milesian ideas of the previous generation.

The Pythagorean idea that the world can be described in mathematical terms will be taken up, expanded, and

championed by Plato, who makes it one of the pillars of
his theory of Truth. Plato hews closely to Pythagoras and
sees the structured world as written with the language of
mathematics, which for the Greeks meant geometry.
One tradition, contested but handed down by Simplicius, has it that Plato had engraved on the door of his
school, the Academy,

Αγεωμέτρητος μηδεὶς εἰσίτω

No one ignorant of geometry can enter here.[9]

Histories of philosophy often emphasize "antiscien-
tific" aspects of Plato (his critique of explanations in
terms of efficient causes, his preference for a priori
thinking over observation), but Plato also made
immensely important contributions to the development
of science.

In the *Timaeus*, he makes a concrete attempt to realize
his program of describing the world geometrically, by
interpreting the atoms of Democritus and Leucippus,
and Empedocles's four elements, in terms of simple geometric figures. From a scientific point of view, Plato's
results are null, but he is moving in the right direction: it
is by means of mathematics that the physical world will
be efficiently described.

A posteriori, Plato's mistake in this brave attempt to
use geometry to achieve a quantitative ordering of the
world is to neglect to take time into account. He seeks to
give a mathematical description of static atomic forms.
He misses the point that it is the evolution of things over
time that can be rendered mathematically. The laws that
will be discovered subsequently are not spatial geometrical laws but laws describing relations between time and
position—laws that describe what is happening "in conformity with the order of Time." One might say, with a

bit of exaggeration, that Plato would have done well to make a better study of Anaximander.

The young Kepler will make the same mistake in his first beautiful and utterly mistaken attempt to use the proportions of the Platonic solids to account for the dimensions of the planets' orbits, deduced by Copernicus. After studying Copernicus's work in depth, Kepler will be able to correct his own error and discover the three laws that govern planetary motion in time ("in conformity with the order of Time"), this time correctly, paving the way for Newton.

Plato never corrected his mistake. But irrespective of his personal failures as a scientist, the influence of his attempt to mathematize the world would be immense. According to Simplicius, it was Plato who posed the question, "Can we account for the strange movement of the planets in the sky in terms of some simple and orderly motion?"[10] This was the fateful question that would give rise to Greek mathematical astronomy and, eventually, Copernicus, Kepler, Newton, and all of modern science. It was Plato who insisted that astronomy could and must become an exact mathematical science. In the Academy, he surrounded himself with the great mathematicians of the day, such as Theaetetus. In the Academy, the great mathematician and astronomer Eudoxos, Plato's friend and follower, would elaborate the first mathematical theory of the solar system.

Twenty centuries later, Galileo's discovery of the first law of terrestrial motion, which gave rise to modern mathematical physics, was directly motivated by his faith in the Platonic and Pythagorean program of seeking a mathematical truth hidden beneath the surface of things. Galileo makes explicit reference to Plato as his source for this idea. One can say that to a considerable degree, all of Western science is a realization of the project developed collectively by Anaximander, Pythagoras, and Plato of seeking out the mathematical laws hidden beneath appearances. But before the idea of mathematical law came to the fore, the idea of a law that governed natural phenomena—something altogether unknown in preceding centuries—was born in Miletus, in all likelihood in Anaximander's thought.

In the centuries to come, the Greeks sought and identified many such laws. In particular, they discovered the mathematical laws that guide planetary motion in the heavens. Galileo would discover the mathematical laws of terrestrial motion. Newton would show that the laws of terrestrial and heavenly motion are the same, founding modern physics.

It is a long journey and a great adventure, begun when Anaximander put forth his idea that such laws exist and govern the world by necessity. The laws of Galileo and Newton, which form the basis of all modern technology, show how physical variables change "according to necessity" and "in conformity with the order of Time."

SIX

REBELLION BECOMES VIRTUE

Thales is traditionally considered one of the Seven Sages of ancient Greece—more or less historical figures recognized and honored by the Greeks as founders of their thought and institutions. Another of the Seven Sages is Solon, contemporary of Thales and Anaximander, the writer of the first democratic constitution of Athens. According to the traditionally accepted dates, Anaximander is only eleven years younger than his illustrious fellow citizen Thales.

We know nothing of the relationship Thales and Anaximander may have had. We don't even know whether the speculations of thinkers such as Anaximander and Thales were private, or whether there was a school in Miletus along the lines of Plato's Academy and Aristotle's Lyceum. These schools brought together teachers and young students, and their activities included public discussions, lessons, and lectures. Texts from the fifth century BCE tell of public debates among philosophers. Did such debates already exist in Miletus of the sixth century?

As I will discuss later, the sixth century in Greece was the first time in human history that the ability to read and write moved beyond a limited circle of professional scribes and became widespread in large sectors of the general population—essentially the entire aristocracy. Any elementary-school pupil today knows that learning to read and write isn't easy. It must have been even more challenging during the first centuries of the use of phonetic alphabets, when writing was far less widespread than today. Someone needed to teach young Greeks how to write and read. I therefore think it is fair to believe that teachers, tutors, and schools must have existed in the major Greek cities of the time, though I have found no confirmation of this point. The combination of teaching and intellectual research that characterizes the philosophical schools of ancient Athens as well as the universities of today might very well have already been established in the sixth century BCE. In other words, I think it is reasonable to suppose that a true school existed in Miletus.

Whether or not such a school existed, it is nonetheless clear that Anaximander's great theoretical speculations were born of and based on Thales's work. The questions they ponder are identical: the search for the *arche*, the form of the cosmos, naturalistic explanations for earthquakes and other phenomena, and so forth. Thales's influence is clear even in smaller details. Anaximander's Earth is a disk, just like Thales's. The intellectual relationship between Thales and Anaximander is very close. Thales's reflections nourish and give rise to Anaximander's theories. Thales is Anaximander's teacher—figuratively and probably also literally.

It is important to consider in depth this close relationship of intellectual paternity between Thales and Anaximander, because it represents, in my view, perhaps the
most important keystone of Anaximander's contribution
to the history of culture.

The ancient world teemed with masters and their
great disciples: Confucius and Mencius, Moses and
Joshua and the prophets, Jesus and Paul of Tarsus,
Buddha and Kaundinya. But the relationship between
Thales and Anaximander was profoundly different from
these. Mencius enriched and studied in depth Confucius's thought but took care never to cast doubt upon
his master's affirmations. Paul established the theoretical
basis for Christianity, far beyond what is in the Gospels,
but never criticized nor openly questioned the sayings of
Jesus. The prophets deepened the description of Yahweh
and of the relationship between him and his people, but
they most certainly did not start from an analysis of
Moses's errors.

Anaximander did something profoundly new. He
immersed himself in Thales's problems, and he
embraced Thales's finest insights, way of thinking, and
intellectual conquests. But at the same time he under-
took a frontal critique of the master's assertions. Thales
says the world is made of water. Not true, says
Anaximander. Thales says the Earth is floating on water.
Not so, says Anaximander. Thales says earthquakes are
attributable to the oscillation of the Earth's disk in the
ocean upon which it floats. Not so, says Anaximander:
they are due to the Earth's splitting open. And he goes on
from there. A still-perplexed Cicero, centuries later,
remarks, "Thales holds that all things are made of water.
. . . But of this he did not persuade Anaximander, though
he was his countryman and companion."[1]

Criticism was not absent from the ancient world—far from it. Take the Bible, for example, in which the religious thought of Babylonia is harshly criticized: Marduk is a "false god," his "diabolical" priests are to be stabbed, and so forth. But between criticism and adherence to a master's teaching there was no middle ground. Even in the generations following Anaximander, the great Pythagorean school—decidedly more archaic than Anaximander in this regard—flourished in reverence to Pythagoras's ideas, which could not be questioned. *Ipse dixit* ("He himself said") is an expression that referred originally to Pythagoras, meaning that if Pythagoras had made an assertion, it must be true.

Halfway between the absolute reverence of the Pythagoreans for Pythagoras, of Mencius for Confucius, of Paul for Jesus, and the rejection of those who hold different views, Anaximander discovered a third way.

Anaximander's reverence for Thales is manifest, and it is clear that Anaximander leans heavily upon Thales's intellectual accomplishments. Still, he does not hesitate to say that Thales is mistaken about this or that matter, or that it is possible to do better. Neither Mencius nor Paul nor the Pythagoreans understood that this narrow third way is the most extraordinary key for the development of knowledge.

In my view, modern science in its entirety is the result of the discovery of this third way. The very possibility of conceiving it can come only from a sophisticated theory of knowledge, according to which truth is accessible, but only gradually, by means of successive refinements. Truth is veiled but may be approached by means of a sustained, almost devotional practice of observation, discussion, and reasoning, where mistakes are always possible. The practice of Plato's Academy is obviously based on this idea.

The same is true for Aristotle and his Lyceum. All of Alexandrian astronomy grows out of the continuous questioning of the assumptions made by earlier masters.*

Anaximander was the first to pursue this third way. He was the first thinker able to conceive and put into practice what is now the fundamental methodological credo of modern scientists: make a thorough study of the masters, come to understand their intellectual achievements, and make these achievements their own. Then, on the basis of the knowledge so acquired, identify the errors in the masters' thinking, correct them, and in so doing improve our understanding of the world.

Consider the great scientists of the modern era. Isn't this precisely what they did? Copernicus did not simply awaken one fine day and proclaim that the Sun was at the center of our planetary system. He did not declare that the Ptolemaic system was an illustrious bit of nonsense.† If he had, he would never have been able to construct a new, effective mathematical representation of the solar system. No one would have believed him, and the Copernican revolution never would have occurred. Instead, Copernicus was thunderstruck by the beauty of the knowledge reached by Alexandrian astronomy and summarized in Ptolemy's *Almagest*, and he immersed himself fully in its study. He appropriated Ptolemy's

*The commonly held belief that Ptolemy's astronomy was in thrall to Aristotle's physics is profoundly untrue. Ptolemy's primary and specific technical contribution, for example, is the equant, which flagrantly violates Aristotelian (and Platonic) principles of motion: Ptolemy's planets do not travel on their rings at a constant velocity, as Aristotelian physics requires.

†Alas, this is precisely how Copernicus's discoveries are presented in today's schoolbooks.

methods and recognized their efficacy; it was in this way, by exploring the nooks and crannies of Ptolemy's work, that he came to recognize its limits and find ways to radically improve it. Copernicus is very much a son of Ptolemy: his treatise, *De revolutionibus orbium cœlestium*, is extremely similar in form and language to Ptolemy's *Almagest*, so much so that one can almost call *De revolutionibus* a revised edition of the *Almagest*. Ptolemy is unquestionably Copernicus's master, from whom he learns everything that he knows and is useful to him. But to move forward, Copernicus must declare that Ptolemy is mistaken—not just in some detail, but in the most fundamental and seemingly best-argued assumptions of the *Almagest*. It is simply not true, as Ptolemy maintains in an ample and convincing discussion at the beginning of the *Almagest*, that the Earth is immobile and at the center of the universe.

Einstein and Newton have precisely the same relationship. And this is not just true for the great scientists: countless scientific articles of today have precisely the same relations to the works preceding them. The strength of scientific thinking derives from the continuous questioning of the hypotheses and results obtained in the past—a questioning that, just the same, takes as its point of departure a profound recognition of the knowledge value contained in these past results.

This is a delicate balance to strike, one that is anything but obvious and natural. In fact, as I have observed, it is unknown in all of the human speculation that has come down to us from the first millennia of recorded history. This delicate process—following and developing the master's path by criticizing the master—has a precise beginning in the history of human thought: the position that Anaximander assumed vis-à-vis his master Thales.

Anaximander's novel approach immediately inspired others. Anaximenes, a few years younger than Anaximander, picked up on the idea and, as we have seen, proposed a modified and much richer theory of the *arche*, or first cause. Once the path of criticism was open, it could not be closed. Heraclitus, Anaxagoras, Empedocles, Leucippus, Democritus: not one of these thinkers hesitated to speak his mind on the nature of worldly things. Only to an inattentive observer can this variety of viewpoints and crescendo of reciprocal criticisms seem a growing cacophony. Instead, it is the triumphant beginning of scientific thought, the earliest exploration of the possible forms that thinking about the world can take. It is the start of the road that has given us everything, or nearly everything, that we study and know about the world today.

According to a classic thesis, a scientific revolution comparable to the one in the West did not take place in Chinese civilization—despite the fact that for centuries Chinese civilization was in many ways broadly superior to the West—precisely because the master in Chinese culture was never criticized or questioned.[2] Chinese thinking grew by elaborating on and developing established knowledge, never by questioning it. This seems to be a reasonable thesis because one can otherwise barely fathom the fact that Chinese civilization, so overwhelmingly great, never managed to understand that the Earth was round until the Jesuits arrived and told the Chinese so. In China, it seems, there was never an Anaximander.

Or, as we will see in the next chapter, if there was one, the emperor probably had him beheaded.

Figure 14. A splendid Mycenaean fresco from the eighth century BCE, known as The Lady of Mycenae. It shows a goddess receiving a gift offering. (*The National Archaeological Museum of Athens*)

SEVEN

WRITING, DEMOCRACY, AND CULTURAL CROSSBREEDING

In earlier chapters, I have argued that an important part of the methodology of scientific thinking has its origins in the Milesian school, in particular with Anaximander. Naturalism, the first use of theoretical terms, the idea of a natural law that necessarily determines how events unfold in time—all of this is Milesian in origin. Above all, Miletus gave the world the combination of elaboration and criticism within an intellectual line of inquiry, and the general idea that the world may be different from how we perceive it: to better understand the world, it may be necessary to radically revise our image of it.

It is surprising that, in the history of the world, all of these momentous intellectual steps happened at the same time and rather suddenly. Why at that moment, why the sixth century BCE? Why in Greece, why Miletus? It is not hard to find some possible answers to these questions.

A rich civilization had flourished in Greece nearly one thousand years before Anaximander, in particular between the sixteenth and twelfth centuries BCE, in centers such as Mycenae, Argo, Tiryns, and Knossos. This is roughly the period of which the *Iliad* sings, though the *Iliad* itself was composed much later. In the memory of the Greek people, this period would remain a fabulous era of splendor. This civilization is known today as Mycenaean or, more correctly, Aegean. Mycenae was the first city discovered during modern archaeological digs, but it does not seem to have been a major center. Its civilization left us traces of grand palaces, rich tombs, gorgeous frescoes (figure 14), and elaborate manufactured goods.

Starting in 1450 BCE, Mycenae dominated Crete, itself the cradle of an ancient, thousand-year-old civilization. Throughout the fourteenth and thirteenth centuries, Mycenae's sphere of influence continued to expand, and the Greeks came to dominate the western Mediterranean, a region in which the Cretans had once prevailed. Mycenae would conquer Rhodes, Cyprus, Lesbos, Troy, and Miletus. Its conquests extended to Phoenicia, Byblos, and Palestine.

From the Cretans, Mycenaean civilization inherited the use of writing: a script known as Linear B (figure 15), which was radically different from classical Greek.

The deciphering of Linear B in the 1950s opened a window onto the world of Mycenae. An unexpected image emerged, that of a civilization with social and political structures much closer to those of Mesopotamia than the ones that would emerge in Greece centuries later.

Mycenaean society was organized around great palaces, where the sovereign and his court lived. The sovereign was considered divine or semidivine, the inter-

Figure 15. Tablets from the thirteenth century BCE showing Linear B script. The one on the right has information about an order of wool. (*The National Archaeological Museum of Athens*)

mediary between the gods and society. All elements of power and sovereignty, both political and religious, were concentrated in his person. The court was the political, economic, religious, administrative, and military center of society, where wealth and power were amassed. All of the territory's production passed through the court, which was also the center of trade—trade that could reach very long distances: golden objects of Mycenaean craft have been discovered as far away as Ireland.

The court had at its disposal a highly organized administration in which writing, practiced by profession-al scribes, played a key role. Their archives record every-thing that touched on the production of crops, livestock, trades, how much individuals had to pay the court in resources and goods, slaves (both private slaves and the ones belonging to the king), the taxes imposed by the palace on individuals and groups, the number of men that each village was required to enroll in the army, etc.[1] Such a structure left little room for individual initiative. All exchanges passed through the palace, which was the center of the network. This was precisely the political and social structure of the Mesopotamian world.

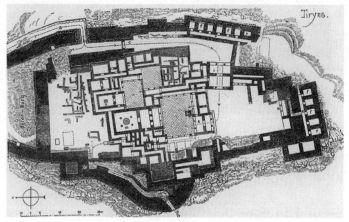

Figure 16. A map of the palace of Tiryns.

The Mycenaean world crumbled around 1000 BCE for reasons that remain unclear. The traditional explanation involves the invasion of the Dorians. There followed several centuries known as the Greek middle ages, from which almost nothing survives. We have no sites, practically no hand-crafted goods, and no writing from this period. Trade seems to have ceased; living conditions must have regressed dramatically.

It is possible that the economic and social difficulties of this period triggered or strengthened emigration from Greece and the establishment of colonies in Asia Minor, along the Black Sea, in Italy, and elsewhere.

Greece emerged from its "dark ages" in the eighth and seventh centuries BCE, the two centuries before Anaximander. Phoenician trade reestablished contacts between the Greek world and the East, contacts that had ceased when the Mycenaean empire fell. Greece once again became prosperous; trade took off again and became increasingly lively; population grew. Agriculture evolved from survival crops, especially grains, to crops for trade, especially olive trees and vineyards. The network of colonies and the trade it encouraged quickly became a source of wealth. Archaeological evidence from this era suddenly becomes more abundant.

Written documents reappeared, but the writing was no longer the Linear B of the Mycenaean era. It was a completely new style of writing, based on an alphabet that the Greeks inherited from the Phoenicians.

The Greek Alphabet

Newly reborn trade had brought the Greeks in close contact with the Phoenicians, who had long dominated naval trade in the Mediterranean. Thanks to these exchanges, the Greeks learned how to use the Phoenician alphabet, which they adapted to their own language. This adaptation brought about a change whose importance cannot be overstated.

While the Greek and Phoenician alphabets look similar, they are not. Both consist of roughly thirty letters. The way the two alphabets work, however, is radically different. The Phoenician alphabet is consonantal; in a given word, only the consonants are written. The preceding sentence, for example, would look more or less like this if I were using a consonantal alphabet: "th phncn lphbt s cnsnntl n gvn wrd nl th cnsnnts r wrttn."

In order to read such an alphabet, one must already possess a fairly clear idea of the subject under discussion and be able to recognize consonant clusters as words. This system can work well in some limited areas—accounting, or the transcription of trade negotiations—but it isn't very versatile in a general context.

A consonantal alphabet may seem to be a somewhat paradoxical idea (why not include vowels?), but its invention represented an immense step forward with respect to earlier forms of writing that had been practiced for thousands of years, such as cuneiform script, used since the fourth century BCE in the area around Mesopotamia; hieroglyphs, which emerged in Egypt a bit later; and Mycenaean Greek's Linear B.

Though they include some phonetic elements, cuneiform and hieroglyphic writing use hundreds of different symbols. In practice, one has to already know a word's written form to be able to write it or recognize it in a text. This requires vast expertise, which in turn requires long apprenticeship. Because of this, writing remained the domain of professional scribes for millennia. Ancient sovereigns and princes did not know how to read and write.*

The Phoenician consonantal alphabet, probably conceived to meet the need for efficiency and flexibility among a people of merchants, simplifies writing drastically. Some thirty letters do the work of hundreds of symbols. The combinations of these letters, determined by the consonantal sounds within each word, codify words in an astute and efficient manner. But learning how to reconstruct words based on their consonants alone continues to requires a high level of expertise. The training needed to master it makes it a skill still reserved to the few.

---

*Hammurabi may have been an exception to this norm, and an important one. Many of his messages seem to have been written by the same hand, and it has been suggested that he himself wrote them. Thousands of years later, Charlemagne did not know how to read and write either.

Around 750 BCE, little more than a century before Anaximander's birth, the Greeks adopted the Phoenician alphabet. But in so doing they stumbled upon a crucial detail—Indo-European phonetics is simpler than Semitic phonetics: Greek has fewer consonants than Phoenician. Similarly, English, an Indo-European language, has fewer consonants than Arabic, a Semitic language with many guttural sounds. Therefore, in adapting the Phoenician alphabet to Greek, some Phoenician letters are left over, the ones corresponding to consonantal sounds that do not exist in Greek. These characters are α, ε, ι, ο, υ, and ω.

Someone in Greece had an idea: use these leftover letters to indicate the sound of the vowels. This way, the many vocalic inflections of the same consonant—*ba*, *be*, *bi*, *bo*, and so forth—all rendered in Phoenician with the single letter β, could be distinguished from one another as βα, βε, βι, βο, etc. It may seem a small idea, but it was a global revolution.

This indeed was the birth of the first complete phonetic alphabet in human history. Compared with the difficulties of previous centuries, reading and writing became almost child's play. With the phonetic alphabet, one could simply learn to listen carefully to each syllable and decode its component consonant and vowel sounds. Conversely, one could learn to sound out the sequence of written letters—*b*, *a*, "*ba!*"—as we all did in elementary school in order for a text to literally begin speaking to us. Instead of recognizing the written word, one could simply pronounce it and recognize it by the sound, even without preliminary knowledge of the particular written word in the text.

The first technology in human history capable of preserving a copy of the human voice was born.

Why did this relatively simple reform in writing need to wait for the Greeks? Couldn't anyone else have accomplished it in the previous four thousand years writing had been in use? Isn't it completely obvious that phonetic writing is a good idea?

I don't have answers to these questions, but the following considerations may be relevant. If a phonetic alphabet is so obviously rational, why is it that, say, France, England, the United States, and China continue to use writing systems that so flagrantly violate the norms of phonetic writing? In French, for example, the word for "water" is spelled "e-a-u" but pronounced as "o." In China and Japan, despite the introduction of new phonetic writing systems (pinyin and kana, respectively), the old characters that have very few phonetic elements remain largely dominant. Rigidity prevails over reasonableness in human cultures. Perhaps it took a new people without writing to start over on a more sensible basis.

Or perhaps it took precisely a people who had known writing five centuries earlier and lost the ability to write but kept the memory of it. As a result, this people could perhaps recognize the value of neighboring peoples' writing systems without being subject to a sense of reverence toward a mysterious, exotic, and arcane technology.

An intelligent merchant or politician in Greece could probably still see ancient Mycenaean inscriptions on the ruins that remained of the ancient civilization sung by the *Iliad*, ruins that might not have disappeared entirely during the seven centuries. He knew that his ancestors in that ancient time of splendor knew how to write. When he came into contact with Phoenician scribes, this man would have grasped the usefulness and advantages of regaining this type of technology—without, however, feeling obliged to copy it slavishly in every detail.

The adaptation of the Phoenician alphabet into the Greek alphabet is so rational and well conceived that I believe it must be the result not of a chance transformation but, instead, of an artificial construct. Rarely does natural evolution lead to structures that lack exceptions and incongruities. I suspect, then, that the use of the Greek alphabet could have been artificially constructed sitting at a table, taking the Phoenician alphabet as the point of departure. To my knowledge, the only other language that uses a perfectly phonetic language is Esperanto, an artificially constructed language. Even in the later classical epoch, Athens passed laws governing the use of the letter ?.

Whatever the turn of events, starting in the mid-seventh century, Greek civilization, which was in its early, vibrant, developing phase, had a true phonetic alphabet at its disposal.

In ancient societies, writing was the exclusive domain of scribes, and knowledge tied to writing was often kept jealously secret. Here, for example, is the text of a cuneiform tablet from Nineveh known as "On Secret Knowledge" (figure 17):

> Secret tablet of Heaven, exclusive knowledge of the great gods, not for distribution! He may teach it to the son he loves.
>
> To teach it to a scribe from Babylon or a scribe from Borsippa or any other scholar is an abomination to Nabu and Nisaba.
>
> Nabu and Nisaba will not confirm him as a teacher. They will condemn him to poverty and indigence, and they will kill him with dropsy![2]

Why would scribes choose to spread knowledge, simplify writing, and then render themselves unemployable? Why would kings choose to turn writing into the her-

Figure 17. Tablet known as "On Secret Knowledge." (*British Museum*)

itage of all? So that they would be driven out, like the
kings of Greece?

To be sure, the idea of secret knowledge did not disappear from the Greek world or in the following centuries. It dominated the Pythagorean school, just as it did
several Alexandrian centers of knowledge, often for military reasons. Marseille, for example, was known in
antiquity for its great secrecy in guarding its military
technologies. The same spirit of secrecy lives on today in
the jealousy with which the United States Department of
Defense guards the results of its scientific research. But
in Greece, with no scribes, no great sovereigns, no
palaces, and no large priestly castes, a new form of
knowledge was born—one not kept secret but rather
proudly broadcast to all.

An immense cultural distance separates the secrecy of
the cuneiform tablet shown earlier and its modern-day
disciples in the Pentagon from the openness of
Anaximander, who set science in motion by entrusting
the entirety of his knowledge to a treatise in prose that
anyone could read—so that anyone could make it his
own, and criticize it, as he himself had done with Thales.

In the seventh and sixth centuries BCE in Greece, for the first time in the history of the world, writing became accessible to many. Knowledge was no longer the exclusive heritage of a closed confraternity of scribes: it became a heritage shared by a large ruling class. Shortly thereafter came the immortal words of Sappho, Sophocles, and Plato.

SCIENCE AND DEMOCRACY

O gentlemen, the time of life is short! . . .
And if we live, we live to tread on kings.
—Shakespeare, Henry IV

Coming out from the Greek middle ages, therefore, was a novel form of civilization, profoundly different from its Mycenaean past. Semidivine kings were no longer there. In the economically and culturally reborn Greece of the seventh century BCE, there was no centralized power, no organized religious authority, no church or powerful priestly caste, no scribes with secret knowledge, no sacred text.

For the first time, the city, the *polis*, was conceived of as an autonomous entity that made decisions on its own behalf. Free discussion and direct participation of citizens settled municipal decisions.

The political structure of these *poleis* was varied and complex. There were kings, then expelled kings, aristocracies, tyrannies, democracies, competing political parties, written constitutions, and rewritten constitutions. Different ways of organizing the *res publica* and its governance were constantly tried out and experimented with. The Greek *poleis* were places where a large class of citizens, many of whom knew how to read and write, discussed among themselves how power should be apportioned and important decisions arrived at. The process of establishing democracy began. Solon, who wrote the first democratic constitution in Athens, was a contemporary of Anaximander's.

Alongside this desacralization and secularization of public life, which passed from the hands of divine kings to those of citizens, came the desacralization and secularization of knowledge. The law that Anaximander sought in order to grasp the cosmos was akin to the law that the citizens of the Greek *polis* sought in order to organize communal life. In both cases, it was no longer a divine law; and in both cases, the law was not handed down once and for all but was instead questioned again and again.

The ancient cosmogonies that shaped foundational myths, from the Babylonian *Enuma Elish* to Hesiod's *Theogony*, mentioned in chapter 1, tell of a world in which the order is established by a great god, Marduk or Zeus, who takes power. Following a long period of battles and confusion, a deity triumphs and establishes an order that is at once cosmic, social, and moral. Hesiod's *Theogony* is a hymn to the glory of Zeus, the founder. This is the mental order of a society that is born from and organized around the figure and power of its sovereign, the prime motor and guarantor of civilization itself.

When Greek cities expelled their kings and discovered that a human collective, even a highly civilized one, needed no king-god to exist—that, in fact, it prospered even more in the absence of a king-god—at this moment, human beings shook off their subjugation to creators and law-giving gods. They began to forge other paths for understanding and ordering the world.

Conceiving a democratic political structure means accepting the notion that the best ideas emerge from the discussions of the many and not from the authority of a single power. It means acknowledging that public criticism can determine the best ideas, and that it is possible to debate and converge at reasonable conclusions. These are the very hypotheses that underlie science's search for knowledge.

The birth of science and the birth of democracy, therefore, have a common foundation: the discovery of the usefulness of criticism and dialogue among equals. In criticizing his master Thales, Anaximander did nothing more than transport onto the plane of knowledge what was already common practice in the *agora*, the places of assembly, in Miletus: not accepting uncritically or reverently the divine or semidivine lord of the day, but instead criticizing the ideas of a citizen magistrate, not out of lack of respect, but out of the shared conviction that a better proposal can always be found.

The old absolute power of kings and priestly castes collapsed, and a new space opened up in which a new culture was born. Men leared to distrust a sovereign's absolute power and the priests' traditional wisdom. Something profoundly new was born in both the structure of society and humanity's quest for knowledge.

The Greeks found their cultural identity in Homeric poetry, which sang of their glorious past. But the Homeric gods were neither fully credible nor truly majestic. As one critic wrote, "There is no poem less religious than the *Iliad*."[3] In this world without a center, without all-powerful gods, a space opened up for a new kind of thought.

There is, therefore, a clear relation between the new social and political structure and the birth of scientific thought.[4] Common elements abound: secularization; the notion that the laws of the ancients, like their ideas, are not necessarily the best; the conviction that the soundest decisions can emerge from a discussion among many and not from sovereign authority or reverence for tradition; the idea that public criticism of ideas is useful for determining the best one; the idea that it is possible to debate and come to conclusions together. These views are at the root of the political process in both ancient Greece and the modern world, and they form the basis of the birth of scientific reflection on the world.

This is, in some sense, the "discovery" of the scientific method. Someone proposes an idea, an explanation. The process doesn't end there. The idea is seriously considered and criticized. Someone proposes a different idea, and comparisons are made. The extraordinary thing is that this process can converge toward agreement. In this way, a group of people can forge a common conviction, or a majority view, or in any event a shared decision.

In the field of knowledge, the discovery is that allowing free criticism and questioning, granting anyone the right to take part in discussions, and taking all proposals seriously does not result in an inconclusive cacophony. On the contrary, it leads to the rejection of the weak hypotheses and the emergence of the best ideas.

It did not last long. A few centuries later, the Roman Empire would once again bring back absolute power to a single ruler, and Christianity would once again bring back knowledge in the hands of the divinity. The marriage of imperial and Christian powers would bring theocracy back.

Still, throughout a large area of the world, for a few centuries human beings were free from theocracy and the dominance of religious thinking. Miletus was independent during Anaximander's lifetime and part of a league of other Ionian cities. This league didn't allow for the dominion of one city over others, but instead opened up a common space for the debate of ideas and decisions of common interest. The building in which the Ionian League met, its parliament, was one of the first parliaments in the history of the world—perhaps the very first one. Precisely when human beings replaced the palaces of divine sovereigns with parliaments, these same human beings took a look at the world around them, freed it from the obfuscation of mythical-religious thought, and began to understand how the world around them functioned. The Earth was no longer a large plate; it became a stone floating in space. Intelligence, like the Earth, was set free to fly.

CULTURAL CROSSBREEDING

Miletus was one of the richest and most flourishing cities of the sixth century BCE, but it certainly was not the only one. Why Miletus? It is probably wise not to look for too precise answers to such questions, but one response comes to mind immediately.

Miletus was the Greek outpost closest to the kingdoms of the Middle East. It maintained close ties with the wealthy kingdom of Lydia, which was at the cutting edge in terms of monetary policy, if nothing else. Miletus traded with the inner Mesopotamian world and had a commercial port in Egypt and colonies along the Black Sea. Ionia's colonies in the western Mediterranean extended to Marseille and beyond. In other words, Miletus was the Greek city most open to the rest of the world, in particular to the influence of the ancient empires and their millennia-old cultures.

Civilizations flourish when they mingle. They decline in isolation. Moments of great cultural ferment always correspond with moments of great cultural encounters. The arrival in Europe of the Arab world's knowledge triggered the Italian Renaissance. The meeting of classical Greek culture and the ancient lore of Egypt, when Alexander the Great drove into the streets of Babylonia and Alexandria, led to the great era of Alexandrian science. Roman poetry flourished when Rome allowed itself to be fertilized by Greek civilization—notwithstanding the boorish, reactionary howls of those who wished to preserve the "purity" of the national cultural identity. Cultural "purity" is invoked even today only by the least-intelligent citizens, terrified by the arrival of peoples different from them.

Writing was born some four thousand years before Anaximander, in Sumer, the cradle of civilization, probably as a result of the encounter of Sumerians with the local Akkadian peoples. The first evidence we have of written language is, in fact, of two languages, Sumerian and Akkadian. Indeed, the oldest cuneiform tablets in our possession include Sumerian-Akkadian dictionaries. Examples of the richness that followed in the wake of the mingling of cultures are innumerable.

These considerations shed light on what seems to me the real sense in which the political organization of the Greek *polis* can be thought of as new. Indo-European tribes or other nomadic peoples did not have the centralized political structure dominated by an absolute, semi-divine king. The sharing of power within assemblies of free men probably existed before the Greek *polis*. It existed several centuries later in the Germanic tribes described by Tacitus, and it is difficult to believe that the assembly of German freemen could be traced back to the influence of the Greek *polis*. The sharing of power among a large group of freemen, therefore, was probably not a Greek discovery. What was new in the Greek *polis* was the meeting of such a structure of shared power with the Mediterranean's cultural riches, accumulated in the palaces of divine monarchs. This encounter taught Greek cities writing, the systematic observation of the heavens, the rudiments of mathematics, the architecture of great temples, and much more. It taught the Greeks a perspective infinitely vaster than those of a tribe of nomadic warriors.

Miletus was where the emerging Greek civilization and the ancient wisdom of the Middle East met. Tradition has it that Thales traveled to Babylonia and to Egypt, where he measured the height of the pyramids. What image better sums up the encounter of Greece's new geometric science with Egypt's ancient traditions? According to Herodotus, Solon set out on his travels θεωριης ἑινεκεν, "out of curiosity."[5] Ancient authors make explicit mention of only two voyages by Anaximander, to Sparta and Apollonia, on the Black Sea. But his thought bears obvious traces of foreign influence, and some scholars today have considered even a Persian influence on his thought.

Plato, too, two centuries after Anaximander, records stories of voyages to Egypt and conversations with Egyptian priests that took place in the time of Solon— that is, during Anaximander's lifetime—for the purpose of learning things that were unknown in Greece. From the cross-pollination between the Mediterranean's vast traditional knowledge and the new cultural policies of the young Greek, Indo-European world there emerged the immense cultural revolution of Miletus.

Herodotus tells a story that captures in a marvelous way this magic moment in human history. He tells of an experience he had during a trip to Egypt—one that, by his account, recreated an analogous experience by Hecataeus:

> Hecataeus the historian was once at Thebes, in Egypt, where he boasted that he descended directly from a god, in sixteen generations. But the priests reacted with him precisely as they also did with me (though I myself did not boast my own lineage): they brought me into the great inner court of the temple and showed me colossal wooden figures. They counted these statues, showing me that they were precisely the number they had previously told me. Custom was that every high priest set up a statue of himself there during his lifetime. Pointing to these and counting, the priests showed me that each high priest succeeded his father. They went through the whole line of figures, from the statue of the man who had most recently died, back to the earliest. Hecataeus had traced his descent and claimed that his sixteenth forefather was a god, but the priests traced a line of descent by counting the statues, and these were three hundred and forty-five. The priests refused to believe that a man could be descended from a god in only sixteen generations; they refused to believe that a man could be born before a god. And all those men whose statues stood there had been good men, but not gods.[6]

Herodotus's decision to narrate this episode in such great detail testifies to the deep impression that the encounter with ancient Egyptian traditions made upon Greek culture. Hecataeus, like all Greeks of his time, believed that the world was less than twenty generations old and boasted of being a close relative to the gods. Along came the Egyptian high priest who took him to

the ancient, dark temple and showed him evidence of
some 345 generations of human civilization. The Greek
world's short past looked ridiculous by comparison. If
this happened to Hecataeus and Herodotus, it probably
happened to many other illustrious Greeks who visited
Egypt, including Thales and, perhaps, Anaximander.
James Shotwell described this beautifully in 1922:

> We might not be far wrong then if we were to date—so
> far as such things can be dated—the decisive awakening
> of that critical, scientific temper which was to produce
> the new science of history from the interview in the dark
> temple-chamber of the priests of Thebes. Yet we must
> remember that it was the Greek visitor and not the
> learned Egyptian priests who applied the lesson. . . . It
> was there that critical thought was born for the western
> world. In them began that bold spirit of investigation
> which became the mark of the Hellenic mind.[7]

Shotwell was writing specifically of the birth of histo-
riography, but his words are equally valid if applied to
the scientific spirit in general.

Almost like Stanley Kubrick's apes before the mono-
lith in *2001: A Space Odyssey*, a Greek, standing before
the Egyptian statues that so spectacularly disproved his
proud vision of the world, perhaps began to think that
our shared certainties could be called into question.

What opens our minds and shows the limits of our
ideas is an encounter with other people, other cultures,
other ideas.

I would note in passing that all of this can serve as a
warning to us today. Each time that we—as a nation, a
group, a continent, or a religion—look inward in celebration of our specific identity, we do nothing but lion-
ize our own limits and sing of our own stupidity. Each

time that we open ourselves to diversity and ponder that which is different from us, we enlarge the richness and intelligence of the human race. A Ministry of National Identity, like those established of late in some Western countries, is nothing more than a ministry of national obtuseness.

EIGHT

WHAT IS SCIENCE?

The science of which I wish to speak was born with nei-
ther the Copernican revolution nor Hellenistic philoso-
phy, but at the moment when Eve plucked the apple. It
is the need to know, part of human nature.
—*Francesca Vidotto*[1]

Did science begin with Anaximander? The question is
poorly put. It depends on what we mean by "science," a
generic term. Depending on whether we give it a broad
or a narrow meaning, we can say that science began with
Newton, Galileo, Archimedes, Hipparchus, Hippoc-
rates, Pythagoras, or Anaximander—or with an astronomer in Babylonia whose name we don't know, or with
the first primate who managed to teach her offspring
what she herself had learned, or with Eve, as in the quo-
tation that opens this chapter. Historically or symbolical-
ly, each of these moments marks humanity's acquisition
of a new, crucial tool for the growth of knowledge.

If by "science" we mean research based on systematic
experimental activities, then it began more or less with
Galileo. If we mean a collection of quantitative observations and theoretical/mathematical models that can

order these observations and give accurate predictions, then the astronomy of Hipparchus and Ptolemy is science.[2] Emphasizing one particular starting point, as I have done with Anaximander, means focusing on a specific aspect of the way we acquire knowledge. It means highlighting specific characteristics of science and thus, implicitly, reflecting on what science is, what the search for knowledge is, and how it works.

What is scientific thinking? What are its limits? What is the reason for its strength? What does it really teach us? What are its characteristics, and how does it compare with other forms of knowledge?

These questions shaped my reflections on Anaximander in preceding chapters. In discussing how Anaximander paved the way for scientific knowledge, I highlighted a certain number of aspects of science itself. Now I shall make these observations more explicit.

THE CRUMBLING OF NINETEENTH-CENTURY ILLUSIONS

A lively debate on the nature of scientific knowledge has taken place during the last century. The work of philosophers of science such as Carnap and Bachelard, Popper and Kuhn, Feyerabend, Lakatos, Quine, van Fraassen, and many others has transformed our understanding of what constitutes scientific activity. To some extent, this reflection was a reaction to a shock: the unexpected collapse of Newtonian physics at the beginning of the twentieth century.

In the nineteenth century, a common joke was that Isaac Newton had not only been one of the most intelligent men in human history, but also the luckiest, because there is only one collection of fundamental natural laws, and Newton had had the good fortune to be the one to discover them. Today we can't help but smile at this notion, because it reveals a serious epistemological error on the part of nineteenth-century thinkers: the idea that good scientific theories are definitive and remain valid until the end of time.

The twentieth century swept away this facile illusion. Highly accurate experiments showed that Newton's theory is mistaken in a very precise sense. The planet Mercury, for example, does not move following Newtonian laws. Albert Einstein, Werner Heisenberg, and their colleagues discovered a new collection of fundamental laws—general relativity and quantum mechanics—that replace Newton's laws and work well in the domains where Newton's theory breaks down, such as accounting for Mercury's orbit, or the behavior of electrons in atoms.

Once burned, twice shy: few people today believe that we now possess definitive scientific laws. It is generally expected that one day Einstein's and Heisenberg's laws will show their limits as well, and will be replaced by better ones.* In fact, the limits of Einstein's and Heisenberg's theories are already emerging. There are subtle incompatibilities between Einstein's theory and Heisenberg's, which make it unreasonable to suppose that we have identified the final, definitive laws of the universe. As a result, research goes on. My own work in theoretical physics is precisely the search for laws that might combine these two theories.

Now, the essential point here is that Einstein's and Heisenberg's theories are not minor corrections to Newton's. The differences go far beyond an adjusted

*Alas, some scientists today still fall into the trap of believing that we possess, or soon will possess, the so-called "final theory" or "theory of everything."

equation, a tidying up, the addition or replacement of a formula. Rather, these new theories constitute a radical rethinking of the world. Newton saw the world as a vast empty space where "particles" move about like pebbles. Einstein understands that such supposedly empty space is in fact a kind of storm-tossed sea. It can fold in on itself, curve, and even (in the case of black holes) shatter. No one had seriously contemplated this possibility before.* For his part, Heisenberg understands that Newton's "particles" are not particles at all but rather bizarre hybrids of particles and waves that run over Faraday lines' webs. In short, over the course of the twentieth century, the world was found to be profoundly different from the way Newton imagined it.

On the one hand, these discoveries confirmed the cognitive strength of science. Like Newton's and Maxwell's theories in their day, these discoveries led quickly to an astonishing development of new technologies that once again radically changed human society. The insights of Faraday and Maxwell brought about radio and communications technology. Einstein's and Heisenberg's led to computers, information technology, atomic energy, and countless other technological advances that have changed our lives.

*German mathematician Carl Friedrich Gauss, perhaps the greatest mathematician of modern times, might have already given serious consideration to the possibility that physical space is curved. Gauss supposedly organized an expedition to verify the hypothesis by measuring the angles of a huge triangle formed by the peaks of three mountains. (In curved space, the sum of the angles of a triangle is not 2π as in flat space). Gauss is said to have kept the undertaking secret for fear of being ridiculed. The reliability of the anecdote is disputed, but whether true or false, it illustrates how strange the idea could appear, a century before Einstein.[3]

But on the other hand, the realization that Newton's picture of the world was false is disconcerting. After Newton, we thought we had understood once and for all the basic structure and functioning of the physical world. We were wrong. The theories of Einstein and Heisenberg themselves will one day likely be proved false. Does this mean that the understanding of the world offered by science cannot be trusted, not even for our best science? What, then, do we really know about the world? What does science teach us about the world?

SCIENCE CANNOT BE REDUCED TO VERIFIABLE PREDICTIONS

To be sure, we can find reliability in science despite these uncertainties. Newton's theory is not less valuable after Einstein. Anyone who needs to calculate the force of wind on a bridge can use either Newton's theory or Einstein's. The difference in results will be exceedingly small and utterly irrelevant for the practical issue of constructing a bridge that will not collapse. Newton's theory, then, is perfectly adequate to this problem and gives us fully trustworthy results (and ones much simpler to use).

Theories have domains of validity determined by the precision with which we observe the world and by the regimes in which the observed phenomena are situated. Newton's theory remains valid and reliable for all objects that move much slower than the speed of light, such as a bridge or the wind. In some ways, Newton's theory is actually strengthened by Einstein's work, because now we know for sure its criteria of applicability. If a calculation based on Newton's equations determines that a roof under construction is too thin and will collapse with the first snowfall, we would be complete fools to dismiss it on the ground that, after Einstein, Newton is not valid anymore.

Because of this kind of certainty, we can happily rely upon science. For example, if we have pneumonia, science tells us that we have a substantial probability of dying if we do nothing, but we are very likely to survive if we take penicillin. This knowledge cannot be questioned: we can be certain that the probability of survival rises significantly with penicillin, whether or not we have any deep understanding of what precisely pneumonia is. The increase in the probability of survival, within established margins of error, is a certain scientific prediction.

We can thus content ourselves to consider a theory interesting only insofar as it gives us predictions valid within a certain realm and within given margins of error. In fact, one might say that generating predictions is the useful and trustworthy part of a theory, the rest being irrelevant baggage.

That is the conclusion of some of today's philosophers of science. Reasonable, but, to me, not convincing. Is the world as Newton describes it, or Einstein, or neither? If neither, is there anything we know about the world? If all we can say is that certain equations are useful for calculating certain physical phenomena within certain margins of error, then we do not leave science any capacity to help us understand the world. Despite our scientific knowledge, the world remains utterly incomprehensible.

The problem with such a reduction of science to verifiable results is that it fails to do justice to the practice of science, the way it actually grows, and above all to the actual use that we make of it, which is the reason why science ultimately interests us. I explain with an example.

What did Copernicus discover? According to the understanding of science just outlined, he discovered nothing at all. Copernicus's predictive system is less accurate, not more, than Ptolemy's.[4] Moreover, today we know that the Sun is not the center of the universe, as Copernicus believed to have discovered.* What, then, is the value of Copernicus's science? From the point of view just outlined, none.

But what good is a viewpoint that tells us that Copernicus discovered nothing? If we take this position, we must conclude that it wasn't Galileo to be right, but rather his accuser, Cardinal Bellarmino, who insisted that the Copernican system was only a system of calculation and not an argument in favor of the fact that the Sun was truly at the center of the solar system, or the Earth just a planet like the others. But had Bellarmino's viewpoint prevailed, we wouldn't have had Newton, nor modern science. And we would still believe that we are at the center of the universe. Something is wrong with reducing science to verifiable predictions.

An understanding of science where the facts that the Sun is at the center of the solar system and the Earth is a planet among the others are to be considered nonscientific is an understanding of science that is showing its limits.

Scientific predictions are important for at least two reasons: they make technical applications possible (calculating whether a roof will collapse without need to wait for a snowfall), and they are our key tool for corroborating (or falsifying) a theory. (Copernicus began to be taken seriously only after Galileo saw the phases of

*We can say that Copernicus understood that the Earth revolved around the Sun and not vice versa. But even this claim, which remains valid under Newton's theory, becomes confused within Einstein's theory of general relativity, in which both Earth and Sun follow geodesic trajectories and neither is a privileged point of reference.

Venus predicted by his system.) But reducing science to a predictive technique is to confuse science with its technical applications, or to mistake a characteristic verification tool with science itself.

Science cannot be reduced to quantitative predictions. It cannot be reduced to calculation techniques, operational protocols, or the hypothesis-deduction method. These are tools, razor sharp and of fundamental importance. They offer relative guarantees, clarity, means for dodging errors, techniques for unmasking mistaken assumptions. But they are only tools and only some of the tools in play in scientific activity. They are at the service of an intellectual activity whose substance is something else.

Numbers, techniques, and predictions are useful for suggesting, testing, confirming, and putting discoveries to use. But there is nothing technical about the content of discoveries. The universe does not revolve around the Earth. All the matter that surrounds us is just made up of protons, electrons, and neutrons. There are one hundred billion galaxies in the universe, and each of them has one hundred billion stars similar to our Sun. Rainwater is water that has evaporated from land and sea. Fourteen billion years ago the universe was squeezed into a fireball. The similarities between parents and children are transmitted by a DNA molecule. Our brains contain nearly a quadrillion synapses that fire when we think. The limitless complexity of chemistry can be reduced to simple electrical forces between protons and electrons. All the living beings upon our planet have a common ancestor, so we are relatives to the ladybugs. These are examples of natural facts revealed by scientific thinking that have profoundly changed our image of the world and ourselves. Their interest and their cognitive consequences are human, direct, and immense—far beyond quantitative predictions.

The confusion between science as a cognitive activity and science as a producer of testable predictions leaves science open to the critique of the dominion of technology. This criticism, rampant in countries such as Germany and Italy, challenges science as a realm of instruments, blind to the real problem that, instead, would be one of ends. But the criticism itself confuses means with ends. To criticize the technical aspects of science is like judging poets based on the type of pen they use. What counts is not the pen used for writing but the poetry that is written. The reason we take interest in an automobile engine is not because it makes wheels turn; it is because it takes us places that we could not reach by foot. The turning wheels are just the mechanism of an instrument that allows us to journey.

EXPLORING FORMS OF THOUGHT ABOUT THE WORLD

In light of these considerations, what is scientific knowledge? The explicit goal of scientific research is not to make correct quantitative predictions; it is to understand how the world works. What does this mean? It means building and developing an image of the world, which is to say a conceptual structure for thinking about the world, effective and consistent with what we know and learn about the world itself.

Science exists because we are very ignorant and we hold a vast number of mistaken assumptions. Science is born of what we don't know ("What's behind that hill?") and challenges what we think we know, but does not stand up to factual proof or reasoned critical analysis. We used to think that Earth was flat, then that it was the center of the universe. We thought that bacteria were born spontaneously from inanimate matter. We thought that Newton's laws were correct. With each new discovery, the world changes before our eyes. We come to know and see it in a different and better way.

Science is looking further, with the awareness that our ideas often prove inadequate the moment we step outside our own little corner of the world. Above all, it is unmasking prejudices, and building and developing novel conceptual tools with which to think more effectively about the world.

Scientific knowledge is the process of continuously modifying and improving our conception of the world, selectively and constantly questioning the assumptions and beliefs on which it is based, searching for modifications that prove to be more effective.

Scientific thinking explores and reshapes the world. It offers us new images of the world. It teaches us how and in what terms to think about it. Science is a continuous quest for the best way to think about and look upon the world. It is, above all, an ongoing exploration of new forms of thinking.

Long before being technical, science is visionary. Hipparchus could not have devised his mathematics without Anaximander, who knew nothing of mathematics. Giordano Bruno's infinite universe paves the way for Galileo and Hubble. Einstein wonders what the world would look like if he were astride a ray of light, and tells us in a book for the general public that he sees curved space-time as a giant mollusk.[5] Science dreams of new worlds, to then realize that some of these describe reality more effectively than our preconceived notions. This process of rethinking the world is continuous.

The greatest conceptual revolutions—those by Anaximander, Darwin, Einstein—are only the most spectacular examples. But the way we think about the world today, the way we organize our thoughts about it, is different from that of a Babylonian three thousand years ago. This deep change is the result of the slow accumulation of knowledge brought about by a great number of big and small changes. Some new knowledge is well accepted: we no longer dance to summon rain. Some is only partially so: we know that the universe, in rapid expansion, has existed for fourteen billion years, but not everyone accepts this idea. There are some, stubborn and angry, who believe that it is only six thousand years old. Still other knowledge is accepted in the research community but has not yet become widely known. The structure of space-time revealed by Einstein's theory of relativity, and the nature of matter unveiled by quantum mechanics, reveal a world dramatically different from the one we are used to. It will take time for everybody to become accustomed to these changes, just as it took two centuries for the Copernican revolution to permeate everybody's consciousness. But the world changes and continues to change around us, little by little as we come to understand wider and wider aspects of it. The visionary strength of science is this capacity to see further, topple preconceived notions, and reveal new lands of reality.

This adventure is grounded on the entirety of accumulated knowledge, but its soul is perpetual change. The essence of scientific knowledge is the capacity to avoid clinging to certainties and received worldviews, and instead be prepared to change these, repeatedly if need be, in light of our knowledge, observations, discussions, different ideas, and criticisms. The nature of scientific thought is critical and rebellious. It does not suffer a priori conclusions, reverence, or untouchable truths.

THE EVOLVING WORLDVIEW

The central insight of Karl Popper, the great philosopher of science, is that science is not a collection of verifiable propositions; rather, it is a set of theories that, at best, can be wholly falsified. Popper understood that scientific knowledge is not what we can verify directly, as positivists expected. On the contrary, it is based on theoretical constructs that can be contradicted by empirical observations. We hold valid a theory that offers predictions that are corroborated as long as it has never been contradicted ("falsified") by reality. This is not a guarantee that, sooner or later, contradictions will not emerge. If they do, the scientist will press forward in search of a better theory. Scientific knowledge, then, is intrinsically global and provisional and in constant evolution.

The evolutionary nature of scientific knowledge has been investigated in detail by Thomas S. Kuhn.* For Kuhn, a scientific theory is a conceptual framework, a "paradigm," for describing a series of phenomena. Within this paradigm, we can explain experimental data, formulate with precision the problems that we see in the world, and work to resolve them. Paradigms can enter a crisis if they are falsified by experience—that is, if in the course of an experiment, we realize that matters are not unfolding as the theory would predict. More realistically, paradigms undergo crises as empirical data accumulate that they have difficulty neatly accounting for.

---

*The historical and evolutionary nature of scientific knowledge has also been emphasized by the Italian philosophy of science from Federigo Enriques and Bruno de Finetti to Ludovico Geymonat, in the wake of the historicism traditionally dominant in Italy, whether deriving from Marx or Benedetto Croce. Italian philosophy of science, however, has been unable to make itself heard in the rest of Europe and overseas.

When such a crisis emerges, an alternative theory may appear that better accounts for data and phenomena. The new theory may topple the old one and take its place. There can be a strong discord between the conceptual structure of the old and the new theories. The two, in some way, may not speak to each other. As Kuhn describes it, science oscillates between "normal" periods, when there is a dominant theory within which scientists seek to resolve problems, and periods of "scientific revolution," in which the general paradigm is swept away and phenomena are reinterpreted within a new conceptual framework.

This reading of science has been developed in several directions. It has been observed that scientific research is always structured in different schools in competition. Schools tend to die off from stagnation more than anything else, when mounting difficulties guide researchers toward more productive research programs. Major paradigm changes and revolutions are less important, in this perspective. Others have underscored the great methodological variety of the scientific process and emphasized that any attempt to capture its complex evolution within a single logic, as in Kuhn or Popper, hinders understanding more than it helps.

These philosophical studies have clarified many aspects of the real workings of science. Still, as a scientist caught up in this great adventure, I feel that some key points are missing.

Lacking first of all is an awareness of the complex ties that scientific theories have among themselves and with the totality of our knowledge of the world. In the models of the evolution of science mentioned above, theories seem to be independent, isolated structures that can be assembled, used, tossed aside, substituted, and tested, one after the other. It is as if we possessed a fixed and reliable conceptual structure, bestowed upon us by reason, common sense, or "obvious" assumptions about the universe, and this structure allowed us to sift through different scientific theories one after the other.[6]

This model of science is too radical regarding abstract matters and too conservative regarding concrete ones. It is radical because it seems to assume that each new theory is born within a tabula rasa. It is conservative because it does not acknowledge that our most rigid structures of thought are themselves contingent. By taking them as absolute, this model unwittingly props them up, against the very revolutionary nature of scientific thought.

A new scientific theory is never a new structure found in the blue by a scientist's imagination. Rather, it is always a small modification of current knowledge. The human brain does not create *ex nihilo*; it takes one step after another. The bulk of knowledge just passes by from one theory to another. Novelty appears only at the outer edges of our understanding—though such outer edges are sometimes at the root of our beliefs.

Every new scientific theory hinges upon the vast complexity of our worldview and, in turn, every valid theory represents new knowledge, a dynamic element within the evolution of this very same worldview.

Kuhn and, to a greater degree, Feyerabend and Lakatos emphasize the elements of discontinuity in the evolution of science, the conceptual gulf between different theories. Without wishing to diminish the importance of what they have been able to see, I think that they underestimate the cumulative nature of science, which is equally undeniable and plays a critical role, especially in the moments of greatest change. They fail to see that what changes in scientific revolutions is not what could reasonably be expected to change, but instead what no one expected.

One example: Einstein is the modern champion of conceptual novelties and scientific revolutions. When Einstein introduced the theory of special relativity in 1905, he did so in response to a crisis of the kind described by Kuhn: Galilean-Newtonian relativity seemed unable to account for various observational results. In particular, it seemed incompatible with Maxwell's electromagnetic theory, whose effectiveness for describing the world was becoming increasingly obvious at the turn of the century. Seen in a Kuhnian spirit of discontinuity, or according to hypothetical-deductive dogma, the solution to the crisis was to be searched for in the creation of a new theoretical foundation that departed radically from the theoretical assumptions of Galileo, Newton, and Maxwell, or that coincided with their assumptions only insofar as empirical consequences were concerned.

This, however, was in no way what Einstein did. He succeeded on the basis of the opposite strategy: he assumed the gist of Galileo and Newton's relativity—namely the equivalence of inertial systems of reference or the fact that velocity is relative—to be correct. At the same time, he assumed Maxwell's equations and the essential aspect of Maxwell's theory—namely the existence of physical fields—to be correct as well. That is, he took for granted the basic qualitative aspects of the successful theories of the world—the very aspects that, according to Kuhn, are overturned in a scientific revolution! The combination of the two seemingly contradictory assumptions became possible by overturning a third hypothesis—that simultaneity is absolute—and this was sufficient to bring forth a new synthesis, the theory of special relativity. This third hypothesis had been previously assumed tacitly but never made explicit. It was considered inherent to the very notion of temporality and, as such, virtually an a priori condition of thought.

Einstein's revolution, then, was not based on discarding theories and trying out other ones. On the contrary: Einstein took existing theory in dead earnest, and instead discarded something in the a priori conception of the world, something never considered until that moment. Einstein did not play a new game within existing rules of the game; he changed the rules of the game. Time was no longer what seemed so obvious. It was no longer what Kant considered an a priori condition of knowledge. Common sense was overturned—and so much for the Anglo-Saxon reverence for common sense.

Einstein did not discard the qualitative, factual knowledge from previous theories to save the phenomena and the verified predictions. On the contrary: he squarely faced this factual content without reservation. In fact, his act of faith in this factual knowledge was so extreme that he preferred to give up a well-established tenet of common sense: the notion of simultaneity. It was not new experimental data that led to the great conceptual leap represented by special relativity, but rather Einstein's faith in the conceptual appropriateness of theories that had shown themselves to be empirically adequate, their apparent contradictions notwithstanding. This reconstruction of the logic of a scientific revolution stands almost in direct opposition to Kuhn's.

Special relativity is not an isolated example. Copernicus did not abandon Ptolemy's theoretical structure in order to reorganize phenomena in a new direction, driven by new observational data. On the contrary:

thanks to a full immersion in Ptolemaic astronomy, he discovered the conceptual key for a total reordering of the world precisely among the recesses of epicycles and deferents. In his novel constructions, epicycles and deferents are still there, but something seemingly unquestionable is overturned: the stillness of the Earth.

Similar cases abound. Dirac devised the quantum theory of fields and predicted the existence of antimatter on the basis of an almost fanatical faith in special relativity and quantum mechanics. Newton grasped the existence of universal gravity because of his complete faith in Kepler's Third Law and Galileo's realization that motion is governed by acceleration, without the slightest additional empirical input.*

In 1915, Einstein himself, in his most spectacular flash of genius and on the basis of his faith in his own special relativity and Newtonian gravity, discovered that spacetime is curved. In all of these cases, faith in the factual content of previous theories—that qualitative content which some part of contemporary philosophy of science treats as nearly irrelevant—created the conditions for giant leaps forward.

The reality of scientific revolutions is thus more complex than a reorganization of observational data on a new conceptual basis. It is a continuous change at the margins and/or the foundations of our global thinking about the world.

---

*Not all great leaps ahead are of this kind. Others, such as quantum mechanics, Kepler ellipses, Galileo's discovery of the law of falling bodies, or today's standard model of particle physics, are directly read into new empirical observations. In the cases mentioned in the text, the leap ahead was still based on empirical observations (the prime source of scientific knowledge is always empirical), but using the fact that this was already coded and synthesized in the factual content of previous theories.

THE RULES OF THE GAME AND COMMENSURABILITY

Science's leaps forward are most often not solutions to well-established problems. They come from discovering that the problem was ill posed. This is why it is so hard to make sense of scientific evolution as a well-defined problem.

Anaximander did not resolve an open question of Babylonian astronomy. Instead, he reformulated altogether the problem of astronomy. He did not clarify how the heavens move over our heads. He understood that the heavens do not lie just over our heads. Ptolemy did not resolve technical problems within Hipparchus's system by discovering new circles along which planets could travel at constant speed; he postulated that planets move at variable speeds, and never mind Aristotelian physics. Copernicus did not explain the strange coincidences of the Ptolemaic system within the rules of the game of astronomy dictated by Plato: explain the appearance of the heavens in terms of simple planetary movements. He changed the game entirely and set the Earth in motion. Similarly, Darwin resolved a problem that was not a problem at all in nineteenth-century biology, because his contemporaries were convinced that they already knew the answer.

It is so not just for science's great leaps. In a scientist's everyday research, even the most humble and minute undertakings, only rarely is a step forward an answer to a well-formulated problem. More often, a step forward is the realization that the problem must be posed differently in order to be resolved. Students writing their doctoral dissertations under my supervision are often surprised that after three years of work, the content of their thesis is not the solution to the problem posed at the outset. If the problem had been well posed, it wouldn't have taken three years to solve it.

The core of the process that leads to the continuous growing of knowledge is therefore not so much trying new theories and discarding old ones, within a general framework where rules of thinking are clearly posed and the ensemble of empirical data is given. Rather, science is a process that builds continuously upon existing theories—that is, upon existing cumulated knowledge—but continuously revises this knowledge, keeping the possibility open of questioning any aspect of it, including the general rules of thinking that appear to be most certain and beyond question.

It follows that scientific theories are not incommensurable, as some contemporary philosophy of science would have it. Theories can be easily translated into one another, including insufficiencies, approximations, and errors. Copernicus's discovery that the Earth revolves around the Sun remains true within the frameworks of Newton and Einstein. The discovery is translated and re-expressed in the new language. There may be great differences between Copernicus's language and the new ones, but the discovery remains recognizable. In fact, Copernicus's theoretical discovery survives not only as a true fact about nature (the Earth revolves around the Sun) but even as a key conceptual ingredient of the new conceptual systems (there is a "Copernican principle" in Einstein's cosmology).

Perhaps the most obvious example of what I mean is provided precisely by the Copernican revolution itself, the prototype of scientific revolution and conceptual reorganization. Ptolemy's *Almagest* and Copernicus's *De revolutionibus* are two of the finest scientific works ever written. In moving from the first to the second, the cosmos is turned upside down. In Ptolemy, there are Heaven and Earth. One category includes all everyday objects and the Earth upon which we walk, the other includes Moon, Sun, stars, and planets. In Copernicus, there is the Sun in one category; Mercury, Venus, the Earth, Mars, Jupiter, and Saturn in a second one; and the Moon, alone, in yet another category. Before, we were still; after, we are on a top spinning along at thirty kilometers per second. Can one even imagine a greater conceptual leap? Can two so very different conceptual systems even talk to each other?

Well, open the two books: Copernicus's treatise, as observed earlier, is extraordinarily similar to Ptolemy's; indeed, it seems almost a corrected edition of Ptolemy's! Same language, mathematics, epicycles, deferents, tables of trigonometric functions, techniques, same general structure, same meticulousness, and same immense, vast vision. The two are impressively similar, and different from anything else written earlier or later. Incommen-surability? It is obviously the same research program. If there exist two people who truly understand each other, they are Ptolemy and Copernicus. They could almost be lovers.

Science, then, never progresses by starting over from scratch. It moves forward by partial steps. But the small steps may shake its foundations. The mainmast and even a rib can be replaced, but the ship itself is never discarded and rebuilt. We go on patching up the only ship that we have, the ship of our thought about the world—the only instrument with which we can chart a course through the infinity of reality. As centuries pass, the ship becomes unrecognizable. From Anaximander's wheels bearing the stars to Einstein's curved space-time, the keel passes through much water. But no one has ever started over from scratch by devising a wholly new conceptual structure, an entirely new vessel. Why? Because we cannot step outside of our own thought. We think in the terms of the conceptual structure that we happen to have. Thought changes from within, step by step, in the harsh and continuous confrontation with its object: reality. But the space of thinkable thoughts is infinite, and we have explored only an infinitesimal fraction so far. The world stands before us, waiting to be explored.

WHY IS SCIENCE RELIABLE?

Return to the initial question. Why is science trustworthy if it is always changing? If tomorrow we will no longer see the world as either Newton or Einstein have found it to be, why should we take seriously today's scientific description of the world?

The answer is simple: because at any given moment of our history, this description of the world is the best we have. The fact that it can be bettered does not diminish the fact that it is a sharp instrument for the understanding of the world. No one throws away a knife because they think that, someday, a sharper knife must exist.

In fact, the evolutionary nature of science, far from being a source of unreliability, is the very reason for its trustworthiness. Scientific answers are not definitive: they are, almost by definition, the best ones that we have at any given time.

Consider a treatment with herbs from a witch doctor. Can we say this treatment is "scientific"? Yes, if it is proven to be effective, even if we have no idea why it works. In fact, several common medications used today in modern "scientific" medicine have their origin in folk treatments, and we are not sure how they work. This does not imply that folk treatments are generally effective: to the contrary, most of them are not. What makes the difference between "scientific medicine" and "nonscientific medicine" is only the readiness to seriously test a treatment, and be ready to give up prejudices and to change minds if something is shown not to work, or something is shown to work. Exaggerating a bit, one could say that the core of modern medicine is not much more than accurate testing of treatments. A homeopathic doctor is not interested in rigorously testing his remedies: he continues to administer the same remedy, even if an accurate statistical analysis shows the remedy ineffective. He prefers to stick to his theory. This is the opposite of scientific thinking. A research doctor in a modern hospital, on the contrary, must be ready to change his theory if a more effective way of understanding illness, or treating it, becomes available. The treatments of "scientific medicine," like the equations of theoretical physics, are the best available ones at present, the ones that up to now have best passed empirical tests. It is not definitive knowledge, it is not complete knowledge. It is the best available knowledge.

Thus, the reliability of science is not based on the fact that its answers are certain. It is based on the fact that its answers are the best available ones. They are the best available ones because science is a way of thinking in which nothing is considered certain, and therefore remains open to adopt better answers if better ones become available.* In other words, science is the discovery that the secret of knowledge is being open to learning, not believing that we have already tapped into ulti-

---

*Misunderstandings of this argument feed much of today's antiscience. Creationist attacks against Darwinism use the argument that Darwinian theory is being revised, so "science is not sure of its own theses." The confusion is between declaring a theory definitive and stating that one theory is better than another. I do not know if this horse is the fastest

mate truth. The reliability of science is based not on certainty but on a radical lack of certainty.

As John Stuart Mill wrote in *On Liberty* in 1859, "The beliefs which we have most warrant for, have no safeguard to rest on, but a standing invitation to the whole world to prove them unfounded."[8]

Scientific thinking, then, is not so different from everyday thinking. It is the same activity, carried out with more-refined instruments: learning to move about the world by constantly updating our mental schemes. When I arrive in a new city, my idea of the city is approximate at first. I make for myself a simple mental scheme that allows me to get about as much as I need to. If I live in the city, my mental image will grow richer. I may realize that some of my first ideas were mistaken. I will always be able to learn new things about my city. Knowing that, in principle, I may in the future have a better map does not diminish the value of the map that sums up what I know right now. This process of knowledge acquisition is what drives science. In this universe, humanity is like a foreigner just arrived in a new city: we have grasped the basics of how to get around, but there is still so much to learn.

IN PRAISE OF UNCERTAINTY

Realizing that knowledge is provisional moves us further and further away from the dream of so many philosophies: finding a foundation to knowledge that can offer certainty.

animal in the world, but I am sure it runs faster than that donkey. Darwin's ideas might not exhaust everything knowable about the history of life—in fact, they likely don't—but beyond a shadow of a doubt, they are far more adherent to reality than biblical creationism. This we do know for sure.[7]

Francis Bacon and then John Locke grounded the reliability of knowledge in empirical observation, René Descartes in the solidity of "pure" reason. Bacon and Locke's empiricism and Descartes's rationalism played immensely influential roles and opened the doors to modernity. The shattering and liberating impact of their philosophies freed knowledge from the prison of tradition, the sole source of reliability recognized by the Middle Ages, giving free rein to critical thinking. This is the immense legacy of their thought.

But today we have learned that if observation and reason are our best tools toward knowledge, neither guarantees certainty. There are no "pure" facts, observations, or empirical data upon which to found theoretical constructions, because our perceptions are heavily structured by our brains, habits of thought, prejudices, and theories. Nor is there a purely rational procedure of thinking that can grant certainty, because we are never truly able to reset to zero the tangle of our assumptions. If we attempted to do this, we would no longer be able to think.

There is no secure method for avoiding error; in the end, we make mistakes just the same. The very critical thinking that Bacon and Descartes unleashed has shown that observation requires a vast, preexisting conceptual structure, and even reason's most obvious assumptions (Descartes's "clear and distinct ideas") can be mistaken. Observations and assumptions can exist only within an already largely structured system of thinking, one that is a priori riddled with errors. We have no certain point of departure. Our point of departure is always just the ramshackle, error-filled totality of what we think we know.

But uncertainty does not make knowledge worthless. Knowing that the empirical data against which we verify our theories are already laden with theoretical assumptions does not make empirical testing worthless. If our theory is contradicted by experiment, this remains a real fact, solid as rock, even if we do not yet know with clarity where our mistake lies. The fact that the assumptions in our reasoning can be mistaken diminishes not a whit the fact that reasoning is our best cognitive tool.

The persistence of doubt does not diminish the validity of what we know. When I am driving my car, I always do so with a healthy measure of doubt that I could make a mistake; I know well and serenely that I must drive right, toward the bridge, and not left, over the precipice. I trust what I know, but I remain alert to the possibility of making a mistake.

There is no secure, unquestionable basis upon which we can find knowledge. Each time we have deluded ourselves into believing we have discovered the definitive theory of the world, we have played fools. Similarly, each time we have thought we have found the final secret to certainty, the secure point of departure for knowledge, we have later been forced to realize we were wrong.

What, then, is "reality"? The entire history of knowledge has shown us that the world is not as it immediately appears to our eyes: there is something else beyond appearances. Beyond the plain blue sky there is an immense space full of galaxies, black holes, and neutron stars. But the uncertainty of our knowledge and the variability of the scientific pictures of the world tell us that we are not getting to an ultimate picture of reality. Should we therefore think that there is an absolute, unknowable, ultimate reality?

No, because this is an utterly useless notion: if it is unknowable, we know nothing about it, and it makes no

sense to even consider it. Should we then discard the
notion of reality altogether, and take the idealist stance
of reducing everything to thinking? This is equally useless, because our thinking is necessarily thinking "about"
reality. Making reference to something outside ourselves—the world, reality—is part of the structure of our
thinking and our language. What is our knowledge
about, if not reality? Reality is thus not a hypothetical
ultimate unknowable entity, it is that about which we do
learn and know.

And we know quite a bit about it—everything that we
have learned to this point. Reality is that thing that we
know so well but still manages to astonish us, with so
many more aspects yet to discover, and, perhaps, aspects
that we won't discover ever. Reality is not the content of
our thinking; it is what often proves to be quite different
from the content of our thinking. Both in confirming and
contradicting our notions, reality continues to make itself
known. It is this reality that involves and interests us.

The process goes on. Science continues to explore and
propose new worldviews that, slowly but surely, are filtered by experience and criticism. This is happening at
all levels. There are competing research programs, but
every research program is itself composed of competing
research subprograms, and every working day for every
scientist is a competition of microresearch programs
running about in his or her head, prevailing, growing,
going backward, and so forth. The best paths are the
ones that survive. Small and grand theoretical constructions grow and are sometimes capsized. We continue to
explore the limitless, virtually infinite domain of the
thinkable.

In the theory of quantum gravity, my own field of research, time does not exist at the fundamental level. Time acquires meaning only in particular situations. The notion of time emerges only as a result of our ignorance of the state of microphysics. I think the disappearance of time is a necessary consequence of what we have discovered about nature with the theories of Einstein and Heisenberg, if we take these theories seriously, as Einstein took seriously Galileo and Faraday. In a sense this is a very conservative deduction, because it is grounded in believing the factual content of established theories, instead of postulating new theoretical principles. If this deduction is correct, on the other hand, the conceptual leap that we are forced to make in order to combine Einstein and Heisenberg is a truly radical one: giving up the common notion of time at the fundamental level. This step questions the form that Anaximander himself gave to the problem of understanding the world: finding the laws that govern it "in conformity with the order of Time."

In quantum gravity, there may be change, but this change is not ordered along the flow of a single time variable. The laws of the world do not govern the evolution of the world in time ("in conformity with the order of Time"). They govern the relations among the many different quantities describing the world. Only under particular conditions do these relations take the form of evolution over a single time variable.

If this is the case, we must change something even in the rules of the game indicated by Anaximander: we must forget about time as a fundamental structure for organizing our knowledge of the world.

But if we manage to contradict Anaximander on so deep a level, we render him the highest possible honor—the honor of having fully absorbed his greatest lesson: the lesson he gave us in following Thales while at the same time indicating Thales's mistakes.

The rampant antiscientism of our time often attacks an image of science made up of certainty and arrogance, or one of pure numbers and cold technical concerns. This is strange. Few human intellectual activities are as intrinsically conscious of their limits as science, and few burn with such a visionary fire.

With each step, a new world comes into view. The Earth is not at the center of the universe; space-time is curved; we are cousins to ladybugs; the world is not anymore made of up and down, heavens above and Earth below. As Hippolyta says so beautifully in Shakespeare's *A Midsummer Night's Dream* (act 5, sc. 1):

> But all the story [. . .] told over,
> And all their minds transfigur'd so together,
> More witnesseth than fancy's images,
> And grows to something of great consistency;
> But, howsoever, strange and admirable.

It seems to me that humanity's common error is to fear this fluidity and seek absolute certainty—the foundation, the fixed point that would soothe our unease. I think that this attitude is naïve and counterproductive to the quest for knowledge. Science is the human adventure of accepting uncertainty, exploring ways of thinking about the world and being ready to overturn any and all certainties we have possessed to this point. This is among the most beautiful of human adventures.

NINE

BETWEEN CULTURAL RELATIVISM AND ABSOLUTE THOUGHT

The vital paradox of our life and thought is that we act and see only from within a context; yet we cease living and understanding if we cease fighting against the limitations imposed by this context.
—Roberto Unger, *The Self Awakened*

Experience has shown us that aesthetic and ethical judgments, and even truth and falsehood, can differ according to cultural contexts. This fact has made us appreciate the difficulty of evaluating judgments and ideas located in systems of values distant from us, culturally or in time.

Today, this healthy awareness informs many kinds of historical and cultural studies. It helps us shake off some of our natural provincialism. It also protects us (to some extent) from the distorting lens of Western imperialism, whose children we are and whose legacy is the belief that Western power is the natural state of affairs, and the Western point of view is the only reasonable one. This awareness helps us to understand that what for us is true, just, or beautiful is not necessarily so for everyone. If science itself cannot offer certainty, all the more reason, then, not to take as gospel truth that which we happen to consider true.

Unfortunately, however, this sane recognition of the relative nature of systems of value and the contingency of judgments is often taken one step further—to complete relativization of all values, namely to the conclusion that all opinions about truth and falsehood are equally valid, or that all ethical and moral judgments must be considered equivalent. This radical version of cultural relativism is, unfortunately, becoming fashionable among the larger public. I believe that it was born of a deep misunderstanding.

Being aware that we may be wrong is different from claiming that it is senseless to speak of right and wrong. Recognizing diversity and taking seriously ideas that diverge from our own is different from claiming that all ideas are equally worthy. Knowing that a given judgment is born within a complex cultural context and is related to many others does not necessarily imply that we are unable to recognize it is wrong.

The central problem of cultural relativism, understood in this radical sense, is that it is self-contradictory. It is true that there is no notion of truth outside history and culture. It is precisely for this reason, though, that we cannot do without truth values. From what vantage point are they speaking, those who deny the meaning of these values? Are they placing themselves outside of culture in order to preach that it is impossible to exist outside of culture? Outside of history in order to preach that it is impossible to exist outside of history? Aren't they expressing a judgment of value or merit, which, as such, by their own terms, has only a relative value? If so, I prefer to belong to a different cultural framework, where comparisons can be made.

The point is that we are always immersed in a given culture, and it is impossible to step outside. The notion of truth does not exist outside of our universe of discourse, and precisely for this reason we cannot do without the notion of truth. Even if we seek to deny it, we speak always in terms of this notion. It is only from within our domain of discourse that we can speak, make truth claims, and formulate judgments.

This does not imply that we should assume that our own aesthetic, ethical, and truth judgments are absolute, universal, or the best ones. Nor does it imply that we must prefer them a priori to the variants we might find in other cultures, in nature itself, or in the evolution of our own thought. Because it is a structural aspect of humans' linguistic universe that our linguistic worlds are open to exchanges between one another. Different cultures are not separated buckets. They are communicating vessels.

Cultures may differ, but difference does not mean incommunicability. Translation may be difficult and may remain incomplete, but this does not mean that substantial reciprocal influences cannot take place. The fact that we are necessarily part of a given culture does not mean that we cannot communicate with another. On the contrary, dialogue with the other—whether this "other" is nature, a culture different from our own, or a great Egyptian priest who shows us a long line of statues—is the essential characteristic of human discourse. Differences do not stand silently in front of one another. They exert reciprocal influence, give rise to comparisons. By coming face to face, different cultural realities immediately engage one another and modify their own value systems and criteria of truth. Radical cultural relativism is ahistorical foolishness that blinds us to the dialectic of cultures.

Furthermore, differences of judgment among cultures are of the same nature as differences of opinion among groups or single members of a given culture. Indeed, they are of the same nature as the kaleidoscope of thoughts and opinions that floats through our heads when, uncertain, we weigh different options before making a decision. Human thinking is not a series of static, separate cultural hands we are dealt; it is a continuous reshuffling on all levels and all scales. It is an ongoing confrontation with other thoughts and with the "outside world" that we call "reality."

To be sure, we might declare for a moment that all things are the same, that reality is a dream. This allows us to smile like Buddha, but to the extent we decide to go on living in reality, we must get involved, seek to understand, and make decisions. We might choose to do this with a Buddha's smile, but we will always be stepping forward, understanding, and taking a stand.

We believe our truth judgments; we are loyal to our ethical assumptions; and we make choices based on our aesthetic criteria. We do this not by choice or in accordance with an ideology, but simply because thinking and living mean judging and choosing. We do these things from within a system of thought that is rich, variegated, and heterogeneous even within a single culture or our very own heads. Our judgments evolve, grow, meet, and influence one another.

The fact that sacrificing maidens to the gods was once considered good and just does not make it any less reprehensible today. Similarly, awareness of the historic and cultural variability of judgment does not make all choices equal, and does not exempt us from making judgments. It merely makes us more open and intelligent as we evaluate the complexities upon which we judge.

I would like to share an example, one of many that illustrate the confusion that I think reigns on this point. The example concerns precisely the history of scientific thought under discussion in this book.

I recently read a very fine article that compares two similar measurements made by two distant civilizations.[1] The first is the celebrated measurement of the variation with latitude of the height of the Sun on the horizon, performed by Eratosthenes in the third century BCE. The aim of Eratosthenes was to measure the size of the Earth. The value he obtained for the circumference of the Earth is very good—surprisingly close to the one we can find in a geography text today. The second measurement is the same one, made in China at more or less the same time, but with a different goal. Basing their work on a cosmology in which the Earth was flat, Chinese astronomers used the measurement to compute the distance between Earth and Sun, getting to the utterly wrong conclusion that the Sun is very close to the Earth, only a few thousand kilometers above the Earth's surface (figure 18, overleaf).

The article is fascinating and reveals a great deal about analogies and differences between two far-removed worlds and two of our small planet's greatest civilizations. When I finished reading it, though, I was puzzled by one thing left unsaid in the article: that the interpretation of the measurement given by Eratosthenes was correct and led to the fact that the West thereafter always knew the correct form and dimension of the Earth; while

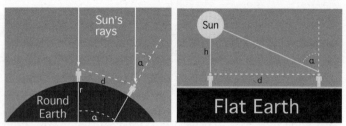

Figure 18. The height of the Sun over the horizon varies with latitude. Left, Eratosthenes's interpretation: the Sun is far away, and the variation is due to the roundness of the Earth. By measuring this variation, we can deduce the radius (r) of the Earth. Right, the Chinese interpretation: the Earth is flat, and the variation is due to the closeness of the Sun to the Earth. By measuring this variation, one can measure the distance (h) of the Sun, which is close by.

the interpretation of the same measurement by Chinese astronomers was wrong and reinforced a fatal error that largely undermined the development of science in China.

I later had the opportunity to meet the article's author, Lisa Raphals, and I asked her what she thought of this difference. She explained that my point of view was mistaken, because the truth value of knowledge about the form of the Earth and the distance of the Sun should be evaluated only within the truth systems of the respective civilizations. It would make no sense to talk of "correct" or "incorrect" in this context.

I do understand that a historian is interested in reconstructing a universe of thinking and its tools, not evaluating it, and Lisa Raphals must write following the rules of the trade. Still, her response reveals something that is deeply disturbing to me in that trade: wouldn't we have a chance to learn more about ancient Chinese versus Greek science if we also asked how did it happen that one got it right and the other got it wrong?

The point here is, of course, whether "right" and "wrong" make any sense at all in this context. I claim they do. Truth values exist within the belief systems of respective civilizations, but this does not make the comparison meaningless.

Indeed, the existence of a difference between the two beliefs is revealed, I believe, by the following historical fact. In the seventeenth century, Jesuit Matteo Ricci brought to China the Greek and European astronomy. The two world visions finally came into direct contact. When Western astronomers came to know of the Chinese calculation, they responded, on the basis of their own belief system, with a smile. As soon as they learned of the Western calculation, Chinese astronomers immediately,* and on the basis of *their* own belief system, changed their worldview, recognizing the Western conception as superior.[2]

This is the difference that the article prefers not to emphasize. The difference proves that Eratosthenes's interpretation was more "correct" than the Chinese astronomers' in a very precise sense—one that Chinese astronomers acknowledged as soon as they could make the comparison. Human systems of value and belief are not impermeable. They communicate, and their dialogue makes clear—not immediately, perhaps, but over time—who is right and who is wrong, and in which sense. Or they are confronted with "factual reality," and this confrontation strengthens one position and weakens another, despite the fact that "factual reality" may be filtered, interpreted, within complex systems of thought. As much as one may wish to believe that the Earth is flat,

---

*This took place long before European colonialism in the Far East. Ricci died in 1610.

the day comes when he must face the fact that Ferdinand Magellan's ship set off toward the West and came back from the East.

To use the comparison between two astronomical measurements to examine similarities and differences between two cultures but choosing to ignore the fatal difference between a correct interpretation and an incorrect one does not lead to a better understanding of similarities and differences between two cultures, in my opinion. Instead, it means that one is willing to ignore something of importance in the history of these cultures.

China today is on its way to becoming again what it had been for most of the fifty centuries during which civilization has existed on Earth: the greatest power on the planet. I don't know if it will complete this transition or what form a future human culture with a central Chinese role will take. Still, I am fairly certain that it will not be a culture in which China will insist on the superiority of the worldview shown in figure 18.

The commingling of our planet's many different cultures is now happening quickly. We are witnessing the birth of a shared civilization, formed by the merging of many different cultures and enriched by the contributions of different traditions. The education of young people in India, China, the United States, Italy, and Brazil is ever more similar and, at the same time, ever more rich and varied. The children of today, both the rich and the poor, are growing up with a breadth of vision of the world incomparably greater than that of their fathers' generation. These cultural encounters also raise irrational fears. Many in the West, for instance, are scared by Arabs, or by China. But if our stupidity does not keep turning everything into conflict and war, the commingling of cultures may continue to grow stronger and open splendid possibilities.

During the few centuries of European colonialism, the West developed an overweening sense of superiority with respect to the other cultures of the world. Until just before World War II, in England and France no less than in Italy and Germany, this sense of superiority took the form of overt racism.* The end of European colonialism and the current weakening of American power have not eliminated, but have tempered, the sense of Western superiority.

As its feeling of superiority weakens, the West begins to doubt itself, the strength of its reason, and the value of its humanism. Taking refuge in an a priori defense of Western superiority is as silly as accepting uncritically the belief that every truth and every value are equal. Recognizing the value of other cultures and leaving behind the naiveté of the West's superiority does not mean disowning the fundamental contributions that all cultures, including those in the West, bring to the world. The West today is learning from the rest of the world (as it always has), but it is also heir to an immense cultural heritage that has vastly contributed and continues to contribute to the rest of the world. In fact, it is still the dominant force shaping the global civilization.

---

*Perhaps the anti-Semitic racism of Fascism and Nazism horrified Europe primarily because its targets were other Europeans, in contrast to the widespread racism of the prewar years, which was directed against non-Europeans. The extent of the German crimes against the Jews during the war is surely one of humankind's most shameful episodes. But the crimes against many non-European peoples, many of whom were utterly exterminated, are not much inferior.

One of the roots of the immense Western cultural heritage is found deep within the culture of classical Greece, where democracy, science, and critical thinking appeared first.

Maurice Godelier wrote "what is born in Greece is not civilization but simply the West."[3] I don't think this is correct. First, the West was not born in Greece: it was born of a mixture of contributions from Greek, Egyptian, Mesopotamian, Gallic, Germanic, Semitic, Arabic, and many other peoples, and developed by many nations, from Rome to England, from France to America, with a continuous influx of external influences. Second, what was born in Greece is not "civilization" either, but it is something that has a universal value in the history of humanity, just as the first African who lit a fire created something that is not "African culture" but a common heritage of all humanity. The Greek heritage spread throughout the Middle East and had an important influence on India and Europe. Modern Europe rediscovered and renewed strands of this heritage, nourished them, and passed them along, with rich original contributions of its own, to the entire world, as America has done and is doing. The fact that this passing along has been mediated by Europe's hateful colonial adventures and today by America's aircraft carriers does not diminish the value of this heritage, something that, curiously, peoples outside the West seem often to understand better than Westerners themselves.

I close this chapter by addressing a view that is at the opposite extreme of cultural relativism: the belief that the only defense against the threat posed by the relativity of values is the restoration of a concept of absolute Truth.

This view is put forward with great insistence in countries in which there is a strong priestly class that exerts a powerful influence on politics, including Iran and Italy, as well as in countries where traditional religious values have increasing weight on politics, such as the United States.

Pope Benedict XVI, for instance, often says that to save ourselves from the relativistic drift, we have to defend the infallible Truth. "Truth," in this context, obviously means the particular truth of those who are speaking. In Italy, it is the Truth in the custody of the Vatican; in Iran, the one in custody of the Ayatollahs; elsewhere, that kept in the Bible, the Qur'an, the Book of Mormon, the Rigveda, or other religious scriptures.

This point of view fails to see that between the certainty of the One Truth and the equal validity of all points of view lies a third way: dialogue and questioning. Human beings often cling to their certainties for fear that their opinions will be proven false. But a certainty that cannot be called into question is not a certainty. Solid certainties are those that survive questioning. In order to accept questioning as the foundation for our voyage toward knowledge, we must be humble enough to accept that today's truth may become tomorrow's falsehood.

Taking this path requires faith in human beings, their being reasonable, and their honesty in searching for truth. This kind of faith in human beings is that of the luminous humanism of the Greek cities in the sixth century BCE at the root of the extraordinary intellectual and cultural flowering of the following centuries, which continues to bear fruit in the contemporary world.

But this faith in humanity is not unchallenged. Many voices rise against it: "Cursed [be] the man that trusteth in man. . . . For he shall be like the heath in the desert, and . . . shall inhabit the parched places in the wilderness, [in] a salt land and not inhabited." (Jeremiah 17:5-6)

The conflict between these two worldviews is ancient. And this takes us to the last topic of this little book.

TEN

CAN WE UNDERSTAND THE WORLD WITHOUT GODS?

If you keep these ideas well in mind,
You will easily see that Nature is free:
Liberated from her superb masters,
She can do all things by herself
Without any need of Gods
—Lucretius, De rerum natura 2

There is one final aspect of the birth of scientific thinking and the revolution initiated by Anaximander that I would like to discuss. The matter is delicate, and I will limit my remarks to reporting some ideas, and to a few observations and questions, in these final chapters.

As emphasized in chapter 3, all texts earlier than Anaximander that are known to us read, structure, interpret, and justify the world in terms of will and actions of the gods. Anaximander gives us something new: a reading of the world where rain is not caused by Zeus but rather by wind and the Sun's heat, and the world is born not by divine decree but from a fireball. The nature of rain, like the origin of the cosmos, becomes an object of a new form of curiosity that leads him to examine natural phenomena in relation to other natural phenomena, with no reference to the sphere of the divine.

In taking this step, Anaximander launches an implicit challenge to religion. As we saw in chapter 3, Anaximander's naturalistic interpretation is global, comprising not only meteorological phenomena but also cosmology, the world's geographic structure, and the nature of life. Such naturalism has a major impact on one of the functions of religion: providing a unifying conceptual structure for understanding the world. This function is implicitly called into question. Anaximander opens up a simple but major problem: are the gods needed to explain the world? For understanding the world, do we need God?

In the ancient sources that have come down to us, there are no indications that Anaximander's text contained any explicit critique of religion. Legend has it that, overjoyed at discovering the solution to a geometric theorem,* Thales sacrifices a bull to Zeus. The problem put on the table by the Ionic school is not a critique of religion or a questioning of the numerous functions that religion plays in human society. The problem of the Ionic school is the comprehensibility of the world, nothing else. This problem is formulated completely excluding the relevance of the divine.

---

*"If A, B, and C are points on a circle where the line AC is a diameter of the circle, then the angle ABC is a right angle."

Some authors, ancient and modern, have questioned this interpretation of the Milesian school and suggested that it is possible to detect a kind of religiosity in the school's approach. Aristotle, for example, writes in *De anima* (411 a7-8), "perhaps Thales came to the opinion that all things are full of gods." I don't believe this interpretation of Thales is correct, at least taken literally. For all of his immense gifts, Aristotle showed little philological rigor in interpreting earlier philosophers. The reliability of Aristotle's account is uncertain and tempered by that "perhaps." Aristotle himself repeatedly criticized the philosophers of the Ionic school, whom he called "physicians," precisely because they sought explanations for all things only in a naturalistic principle, which Aristotle called "physical."

The key point, in any case, is not how Thales and Anaximander conceive of the gods or to what extent their reading of the world partakes of ancient religiosity. The point is that their revolutionary theory for explaining the cosmos is formulated entirely in natural, physical terms. It explicitly and radically excludes all references to the divine and paves the way for all later quests for knowledge that ignore the divine.

On this point we may trust an expert, Saint Augustine:

> [Anaximander] thought that all worlds are subject to a perpetual process of alternate dissolution and regeneration, each one continuing for a longer or shorter period of time, according to the nature of the case; nor did he, any more than Thales, attribute anything to a divine mind in the production of all this activity of things.[1]

"Anything." Saint Augustine was interested in reading the presence of the divine into pagan thought, and did not hesitate to seek out traces of God in the works of the ancient philosophers, wherever he could. If his views on Thales and Anaximander are so drastic, it is clear that there is nothing in their thought that he could recognize as consonant with religion.*

The cultural closeness of Milesian speculation to earlier thought is strong and has often been emphasized. For example, Thales's assumption that all things are made of water carries echoes of Babylonian, biblical, and Homeric cosmology. The very structure of Anaximander's cosmogony hews closely to Hesiod's: same general problem, same structure, similar phases. These relations are natural: Milesian thinking is not born out of nothing, it derives from the culture to which it is grafted. But these similarities must not blind us to the difference— because the difference, in fact, is what counts. Copernicus's treatise is similar to Ptolemy's, but there is a difference, and this difference is Copernicus's achievement. The obvious and immense difference between Thales's and Anaximander's cosmogonies and the earlier ones is the disappearance of the gods. There is no trace of the *Iliad*'s Oceanus, the father of the gods, nor of the god Apsu from the *Enuma Elish*, nor of the God of the Bible who speaks and creates light upon the oceans. There is only water. And so on: an autonomous unfolding of the world replaces the gods' words, struggles, and battles.

*Nicola Abbagnano conveys this point well: "The thesis put forward by modern critics that this philosophy is born of mysticism, from which it supposedly inherited its tendency to consider the physical universe in anthropomorphic terms, is based on arbitrary parallels which have no historical basis. . . . Pre-Socratic philosophers were the first to achieve that reduction of nature to objectivity, which is the first consideration of any scientific examination of nature. This reduction is precisely the opposite of the confusion between nature and humanity typical of ancient mysticism."[2]

Though it contains no explicit questioning of the divine, the entirety of Anaximander's cognitive project is based on the radical stance of ignoring the gods.* This stance unavoidably comes into conflict with the dominant sentiment of the age, grounded in the gods.

A conflict is open. It will have a long and painful history.

THE CONFLICT

The reaction of mythical-religious thought against the new naturalism arose promptly, intensified quickly, and would flare up over and over again, and with violence, throughout history.

Heresy convictions soon began to hit Anaximander's descendants. Anaxagoras was exiled from Athens; Socrates was sentenced to death for the crime of fomenting irreverence for the local gods among youth. This accusation was precisely the one made years before in Aristophanes's *The Clouds*, mentioned in chapter 3, where the scandal is exemplified in terms characteristic of Anaximander: is lightning sent by Zeus, or does it come from a vortex of wind?

Nonetheless, in general terms, the polytheism of the Greek world and of the early Roman Empire, probably already weakened by changing times, managed to coexist

---

*Reading the many ancient fragments that relate Anaximander's explanations for natural phenomena, one feels like asking him, And the gods? And despite the extreme anachronism, one is tempted to imagine the thoughtful, sweet, but slightly clouded face of Anaximander on the bas-relief at the Museo Romano (see frontispiece) looking up from an ancient papyrus, fixing us in silence, smiling, and anticipating the famous answer of Laplace to Napoléon: "Sire, I have no need of that hypothesis."

more or less peacefully with the first flowering of naturalistic thinking. The same cannot be said of the next fifteen hundred years, when monotheism took hold.

The first period of systematic and violent clashes between naturalistic knowledge and religion was the late Roman Empire, immediately upon the arrival of Christianity to power. In 380 CE, Theodosius declared Christianity the empire's state religion. Between 391 and 392, he issued the Theodosian decrees, which imposed religious intolerance. It was a return to theocracy, as in the days of the pharaohs, the Babylonian kings, and Mycenaean civilization. Monotheism was imposed with extreme violence. Philosophical schools were shut down by order of the authorities; ancient centers of learning were destroyed along with pagan temples, which were sacked or transformed into churches, following the pattern inaugurated by Josiah's violently imposing Jewish monotheism centuries earlier in Jerusalem (see chapter 1).[3] Blood was spilled in Areopolis, Greece, in Petra in Arabia, in Raphia and Gaza in Palestine, in Hierakonpolis in Phoenicia, in Apamea in Syria—and, above all, in Alexandria.

Alexandria! It all probably started a short distance from there, about a millennium earlier, at Naucratis, where Greek merchants from Miletus were able to open the first emporium in the land of the great pharaoh, grateful for the aid received from Greek mercenaries in fighting off the Assyrian menace. At Naucratis, the meeting of these Indo-Europeans of free and adventurous spirit with the ancient wisdom of Egypt was perhaps what set off the magic spark that ignited the adventure of knowledge. The heritage of Miletus was taken up by Athens, where the dream of understanding the world by means of human intelligence led to the flowering of

Plato's and Aristotle's schools. Aristotle's young and exuberant pupil Alexander conquered the world, spreading everywhere the splendor of Greek intelligence. The great city that he founded and that bears his name became the center of ancient learning. Ptolemy I—Alexander's friend, his first general, and then first Greek king of Egypt—had Aristotle's famous library transported there from Athens, by a ruse, to then found around it the great institutions of ancient knowledge: the Great Library of Alexandria and the Mouseion. The library had collected texts from all over the world. Every ship that dropped anchor in the port of Alexandria was obliged to deposit in the library all the books it had onboard; death was the penalty for a captain not complying. The books were copied, and the copies were returned to the ship. In the Mouseion, true forerunner of modern universities, the city paid stipends to intellectuals who studied all fields of knowledge.

Half of what we learn today in primary school originated in Alexandria's institutions, from Euclidean geometry to the knowledge of the Earth's dimensions, from optics to the basis of medical anatomy, from statics to the basis of astronomy. Archimedes's letters were addressed to Alexandria. In Alexandria, the precise mathematical laws that govern the motion of the planets in the heavens had been discovered.

The early Roman Empire managed to coexist with proud Alexandria with difficulty. Christian Rome never managed to. The Great Library, the depositary of ancient wisdom, was burned and devastated by Christians.* The pagans barricaded in the great temple

---

*The library was not destroyed centuries later by the Caliph Omar as a telling of events in Christian lands would have it.

of Apollo were stabbed to death. In 415 CE, astronomer and philosopher Hypatia, the woman who probably invented the astrolabe, was lynched by Christian fanatics. Hypatia was the daughter of Theon of Alexandria, the last known member of the Mouseion.

One thousand years after Greek merchants arrived in Naucratis, Christianity seized power and snuffed out this luminous source of knowledge. Reality might have been even worse than this, since we read of these tragic events only in Christian texts—nearly all pagan texts were systematically burned in the following decades. The God of monotheism is a jealous God. More than once this God has attacked and destroyed with blind violence everything that rebelled against him.

The anti-intellectual violence of the Christianized Roman Empire managed to suffocate almost every development of rational thought for many centuries. With Christianity's conquest of the Roman Empire, the absolutism of the ancient empires was restored, but on a vaster scale, and the parenthesis of light and free thinking lit in Miletus one thousand years earlier was closed.

The traces of the great knowledge that had grown out from Anaximander's intellectual audacity remained buried in few ancient codices that survived the fury of the first generations of Christian rulers. These were studied and handed down with almost reverential awe by wise men of India, Persia, and the Islamic world. Mathematicians and astronomers in Islam—including al Battâni, Tusi, Ibn al Shatir—were able to develop aspects of the ancient wisdom. But until Copernicus, no one would be able to make his own Anaximander's fundamental lesson: if you want to truly advance the path of knowledge, you must not just revere your master, study, and build on his teachings. You must seek out his mistakes.

Modern rational thought and modern science, in turn, have clashed repeatedly with religious thought, from Galileo to Darwin, and, on a much vaster scale, from the French Revolution to the Russian Revolution. It is a long, bloody, painful story that need not be retold here, in which violence in the name of religion and against religion repeatedly bloodied Europe.

Following the horror of the great wars of religion that devastated Europe in the seventeenth century, when Europeans massacred each other, each in the name of his own version of the True God, the Enlightenment rebelled against the centrality of religion and left to Europe a beautiful heritage: the idea that different ideas, different faiths, rational and religious thought, might peacefully coexist.

The coexistence of rational and religious thinking that the nineteenth and twentieth centuries inherited from the Enlightenment is based upon a delimitation of spheres, fluid and ambiguous but workable. Religion came to be restricted to ever narrower spheres—for example, to private spirituality; the personal existential motivations of some people ("believers"); a point of reference for ethics and morality, with a continuous renegotiation of the balance between its private and public roles; or the oversight of ritual aspects of events that structure social life, such as weddings and funerals. In the field of knowledge, religion was confined to a possible explanatory basis for very general issues ("Why does the world exist?"), or for issues that are more difficult for naturalistic thought to take on ("What is consciousness?"). This Western model of a division of spheres has been imposed, more or less, all over the modern world as a result of colonialism. We are immersed in it. The American Constitution is grounded on it.

But religious communities often accept this delimitation of spheres only with difficulty, as shown by the Italian Catholic Church's recent political activism, or the prominence of the Religious Right in the United States. The reason for this is clear: this division of spheres in some ways contradicts the very essence of monotheism, which sees itself as the final and complete basis for legitimacy and ultimate guarantor of Truth—and, thus, as the foundation of knowledge. Our civilization today oscillates in this uncertainty about its own foundations: On the one hand, the compromise of democracy, which gives equal a priori dignity to all different points of view. On the other hand, religious thought, which, with some difficulty, can come to accept respectful coexistence with what differs from it, but in Rome and Riyadh, in Washington and Tehran, continues to think of itself as the ultimate repository of truth, the doubting of which is evil.

From a theoretical point of view, the search for a compromise between reason and religion marks the evolution of Christian thought and nourishes the philosophy of the church fathers, from Saint Augustine to Saint Thomas. Seen from the point of view of a modern scientist, these efforts have a kind of desperate, tragic grandeur. Other times they seem to be clutching at straws: great intelligences on a quest for improbable distinctions.

At times they become altogether grotesque. In *The City of God*, Saint Augustine, anxious not to contradict reason, discusses the following problem at length and in great detail. When the dead are resurrected at the end of time, each human being will be reunited with his body, with every last particle of his flesh. Now, if a cannibal has eaten another human being, will the particles of flesh consumed by the cannibal be resurrected as part of the cannibal or of the person who was eaten?* Augustine was undoubtedly a person of great intelligence, and I find it so sad to see such intelligence wasted on issues of this kind.

In the end, the clash between religion and reason is ultimately unsolvable. We have learned to define different spheres of knowledge, and much of ancient and modern science developed peacefully in the bosom of religious sentiment: Thales makes sacrifices to Zeus, Newton explicitly refers to God when introducing his novel notions of space and time. In the minds of many, religious and rational knowledge coexist. There is no contradiction between solving Maxwell's equations and believing that God created heaven and Earth and will judge humanity at the end of time.

But a deeper contradiction remains unresolved and is doomed to return. Conflict is inevitable for two reasons. On a more superficial level, there is the fact that the border between the divine and scientific spheres of competence is always up for debate. But there is a deeper reason: mythic-religious thought relies on the acceptance of absolute truths that must not be questioned, while the very nature of scientific thinking demands that truths, particularly those accepted uncritically, be questioned.

---

*The answer, at the end of a long discussion, is that they will be resurrected as part of the man who was eaten and not as part of the cannibal. If I understand correctly, this is because cannibalism being a sin, the particles of flesh eaten become the cannibal's in fact but not according to law. The cannibal, then, acquires them on Earth, but does not have the right to have them back at the end of days.

Any truce between these two forms of knowledge, however long lasting, is intrinsically unstable.

On the one hand, there is the certainty of having some form of direct access to Truth. On the other hand, there is the recognition of our ignorance, and the continuous challenge to any certainty. Religion, particularly monotheism, may accept critical thinking and a thought in constant evolution only with profound difficulty.

Eve plucked the apple from the tree of knowledge, but for a god posited as the singular and unchallenged God, this was humanity's first sin.

I think that the majority of men and women who make up the many variants of the one civilization in which today's world is immersed believe that a true understanding of the world must include gods or God. In other words, Anaximander hasn't persuaded most of us humans.

This majority believes that this God plays, or at least played, a founding role in the very existence of reality, in the justification of power, and in the establishment of morality and, hence, law. Men and women appeal to God, or to "God's will," for private matters and decisions. The governments of several nations, including Iran, Iraq, and the United States, explicitly invoke God to justify important decisions—for example, if they declare war on one another. Each, of course, is convinced that God is on its side. In my own country, Italy, it is not unusual to read in national newspapers that "without God nothing is understandable." In short, we live in a global civilization where the majority of men and women accept rational scientific thinking as useful and reasonable, but pose gods or God as the ultimate foundation of knowledge.

Other men and women believe that none of this makes sense—that the world is more understandable, or, better, less incomprehensible, when one omits any reference to God or gods. They believe that power must be justified without reference to God, and that there is a basis for morality and law that need not appeal to God. They believe that resorting to divine will as a justification for a nation's momentous decisions is a poisonous relic from a dark past, one that divides instead of uniting us, that has led and leads to war more often than to peace.

In the heart of our civilization, there is a rift on the subject of the role of God or gods. There are extreme positions (from the literal interpretation of the Bible and Qur'an to militant atheism) and a range of intermediate stances that cover a spectrum of compromises and shaded interpretations of what role the gods or God play in society and in our understanding of the world.

The question posited by Anaximander, in other words, is still entirely on the table. The idea of formulating an understanding of the world without reference to the gods was a radical one in the sixth century BCE. It had immense consequences, paving the way for the philosophical and scientific developments that grew, in alternate phases, during the next twenty-six centuries. It represents one of the deepest roots of modernity. But it is not an idea that has prevailed. Many, perhaps most, people in our world dissent.

The naturalistic and scientific-rational approach to understanding the world has won successes that Anaximander could hardly have dreamed of. Greek-Alexandrian science and then modern science made Anaximander's project their own. They extended, completed, and developed it, obtaining as a result a profound and detailed understanding of innumerable aspects of reality—and also, as a result, the entire technology that forms the basis of the modern world and determines our daily life. But the theistic thought that Anaximander put aside remains the most diffuse form of thinking on our planet.

The question raised by Anaximander remains relevant, even burning, today. It divides our civilization. Is it possible to understand reality, the world's complexity, and our own lives, without attributing them to the caprices of gods, or the will of one God?

ELEVEN

PRESCIENTIFIC THOUGHT

But what does it truly mean, Anaximander's bold proposal to conceive the world without recourse to the gods?

What is the essential difference between naturalistic thinking and mythic-religious thinking? Why should seeking to understand nature without reference to the gods be so radically new? Why did human beings before Anaximander always explain the world by means of the gods? In short, what is this mythic-religious thought from which it has been so hard for us humans to distance ourselves? What are the gods?

The question is too vast to be exhausted within this little book, and a full answer is beyond my competence and, I think, beyond what we currently know. But it is central for understanding Anaximander's achievement and the nature of scientific thinking, and in this last chapter I offer some elements of reflection about it. The common definition of naturalism—conceiving the world without recourse to supernatural causes—is a very vague definition, unless we have a precise idea of what these supernatural causes are and, above all, of the reason for their ubiquity. Investigating the nature of the religious reading of the world is necessary in order to understand the thought that defines itself in opposition to it. It makes little sense to speak of an understanding of the world that does without mythical-religious causes without knowing what mythical-religious causes are.

We know little of the history of mythical-religious thought. According to some scholars, humanity's religious or ritual activity goes back at least two hundred thousand years and may even have developed simultaneously with language itself.* At the opposite extreme, some believe that religion emerged during the Neolithic revolution—with the birth of agriculture, demographic growth, and the first stable settlements—only around 8000 BCE.[2] Still, there is a consensus today—based on the many texts that have come down to us and anthropological studies done in the last hundred years on the remaining so-called "primitive" cultures on our planet—that religious thought was the dominant form of thought in all ancient human cultures of which we have evidence.

In *Ritual and Religion in the Making of Humanity*, Roy Rappaport, one of the leading figures in anthropology, argues in detail, based on vast anthropological evidence, that in all known cultures, the sphere of the sacred and divine, in its multifarious forms, represents the universal basis not only for the founding of social, legal, and political legitimacy, but also for the intelligibility and coher-

---

*Archaeological evidence (bear skulls arranged artfully in circles) has been discovered in a Swiss cave that goes back to the Würm glaciations.[1]

ence of thought about the world. When causes are
sought, they are sought in the relationship between the
world of visible phenomena and another world that
underlies, justifies, and guides the visible world. This
other world is made manifest by means of gods, spirits,
demons, ancestors, or heroes who live in a mythic parallel time, or beyond time, or at the beginning of time, or
in other "subterranean" realms that are easily ascribed to
a similar mythic-religious matrix. For thousands of years,
mythical-religious thought was the only form of thought
of which humanity was capable. The universality of religious thought, in its countless variants, is beyond question. Ancient thinking is universally religious thinking.[3]

Given the universality of religious thinking and the
fact that it endures in our own time, it is clear that viewing this thought only as a collection of superstitions and
false beliefs is being blind to something essential tied up
with its strength. What is this strength? The gods were
not merely inventions of human imagination; they represent something basic to the very structuring of humanity's cognitive, social, and psychological experience.
What is this thing? What precisely does Anaximander's
proposal stand in contrast to?

The reasons behind such omnipresence of "another
world," "gods," and the other variants of the divine in
the ancient world is, in my view, one of the most important open questions about the nature and history of
thought.

The Nature of Mythical-Religious Thought

Attempts to answer this question abound and shed light
on some of the facets of a complex whole. As far back as
Democritus, Epicurus, and Lucretius, traditional
motives for religion were sought in the fear of death,

from which everyone supposedly suffers (but is this true?). Other motives include fear of the uncontrollable aspects of the world; aesthetic reverence in the face of nature's immense spectacle; instinctive reaction to the unintelligibility of the world, or to the notion of the infinite; or, finally, a hypothetical, ill-defined innate religious spirituality in individuals.

A classic anthropological interpretation of religion is that of Émile Durkheim. In Durkheim's view, religion functions to structure society, and religious rituals are mechanisms through which solidarity and the spirit of a group are expressed and strengthened ("If the totem is both the symbol of god and of society, are these not one and the same? The god of the clan . . . must therefore be the clan itself,"[4] often condensed in the famous formula "religion is society worshiping itself"). Political power does not use religious power; it is itself religious power. The pharaoh *is* god.

Another interpretation of religion is given by Karl Marx: religion does not express the interests of society as a whole, but only of the ruling class, for whom it is a means of perpetuating the oppression upon the rest of society.

More recent theoretical approaches to the origins of religion and the role it played in the birth of civilization are varied. They range from evolutionary approaches, according to which religion represents a competitive advantage for selected groups or individuals, to radically different hypotheses, in which religion is a parasitic deviation, a useless collateral product of other societal functions.

Some of the most interesting investigations, their hypothetical and controversial aspects notwithstanding, concern the historical evolution of religious thought. In the 1970s, a lively debate was set off by Julian Jaynes's fascinating *The Origin of Consciousness in the Breakdown of the Bicameral Mind*. Jaynes argues against the idea of an ancient origin of divinity, and considers instead the hypothesis that the idea of God was born during the Neolithic revolution, around ten thousand years ago. Groups of humans originally had the form of families guided by a dominant male who commanded the group directly via a personal relationship with each member, precisely as we still see today in the social groups of the great apes. With the Neolithic revolution, some human groups expanded to such an extent that the dominant male could no longer maintain direct contact with each member. "Civilization" is, then, the art of living in cities so large that their inhabitants are not all personally acquainted.

Jaynes argues that the solution that emerged to prevent the scattering of the group was an introjection of the dominant male's figure, whose "voice," ordering what to do, became "heard" by his subjects even in his absence. The voice of the sovereign continues to be heard and revered even after his death. His corpse, preserved for as long as possible, still "speaks," and evolves into the statue of the god, worshiped at the center of every ancient city. The sovereign's home—that is, the home of the god's statue—becomes the temple at the heart of every ancient city.* The system stabilizes over

---

*Archaeological evidence from the earliest urban settlements shows that they were arrayed around the god's home or took the form of a series of nuclei organized around a god's home that contained a statue. This structure is clearly in place in Jericho, at the level corresponding to the seventh millennium BCE; in the Hacilar settlement in Anatolia (7000 BCE); and at Eridu around 5500 BCE, where the god's home is built

the course of millennia and determines the social and psychological structure of ancient civilizations.

In these civilizations, the god was literally the sovereign, the sovereign's father, or the sovereign's ancestor. The gods were the active memory of the sovereigns who were dead but still spoke to humans. The voices of the gods were everywhere, and listening to these voices was the way the members of the ancient civilizations faced stress situations where decision was necessary—as we read, for example, in the *Iliad*. According to Jaynes, humans did not yet possess the complex consciousness of the self in the modern sense: a vast, interior, narrative space in which they portray themselves and the possible consequences of their own actions, in order to be able to articulate complex decisions. Instead, they had introjected a system of rules that reflected social behavioral norms. These manifested themselves as the direct will of the gods. Gods, then, were not imaginative inventions; they were early social man's volition itself.

According to Jaynes, this system underwent a crisis around the first millennium BCE, a period of extremely violent political and social upheavals. It collapsed under the weight of vast population migrations, the development of trade, and the formation of the first multiethnic empires. Given the ever-expanding confusion among different groups, the "voice" of the god, which used to speak to the Homeric heroes, and which Moses and Hammurabi still heard distinctly, became ever more

upon platforms of mudbrick that anticipate the ziggurats. The same can be said for archaeological remains in Mexico and the structures of cities in China and India, which hardly vary at all with respect to this structure. From these structures to Gothic cathedrals the continuity is striking.

remote and eventually was heard only by the priestesses at Delphi, Mohammed, or Catholic saints. The gods drew farther and farther away, into the heavens. Humanity was left alone, at the mercy of a tumultuous world. Jaynes's pages where he describes this period are beautiful. This is the period in which a lament begins to sound, one which we will hear over and over again:

> My god has abandoned me and disappeared.
> My goddess has failed me and stays far away.
> The good angel who walked at my side has
> gone away. . . .[5]

Modern consciousness, in Jaynes's view, is the evolutionary expedient with which to confront this new aloneness: a linguistic narrativization of the self, which became the new means for making complex, articulated decisions when neither the head of the social group nor his introjected voice remained to tell humans what to do. In this interpretation, God is essentially what remains of the memory of the big dominant ape that used to lead the early primates group.

Marcel Gauchet's *The Disenchantment of the World* is another classic theory, from a radically different cultural milieu that, nonetheless, resonates in interesting ways with Jaynes's.[6] Gauchet describes humanity's slow emergence from mythical-religious thought. In his view, religion in the past represented humanity's general economy: it bound together material, social, mental, and (above all) political life. This function, however, exhausted itself over the centuries. Modern states, for the most part, have taken on religion's role in the structuring of the political realm, and religion endures only in tatters—in little more than individual experience and personal systems of belief.

One of Gauchet's interesting ideas is that monotheism does not represent an evolved, superior form of religious thought but, on the contrary, is a phase in the slow dissolution of the centrality and coherence of the ancient religious structuring of thought.

The birth of monotheism relates to the formation of the great empires. The first empires mixed peoples, seized power from the primitive social group (the tribe that identified with its local god), and spawned the idea of a great, distant central power. A single god began to dominate the multiplicity of gods and cults that had once prevailed. In Egypt, Ra, the sun god, emerged as the main god during the fourth dynasty of the Old Kingdom. In Mesopotamia, Marduk, the god of Babylon, came to tower above the legions of other gods as soon as power was centralized in Babylon.

But ancient polytheism was not easily ousted. There were attempts to impose a single god, in particular the epic, dramatic struggle of Amenophis IV, the husband of Nefertiti, who gave himself the name Akhenaten. At the peak of Egyptian imperial glory, Amenophis introduced the monotheistic worship of Aton, centered at Akhetaton. But there was a violent reaction led by the priestly castes. Polytheism was restored upon Amenophis's death and was never really uprooted by the ancient empires. Only the Roman Empire, the West's largest and most stable empire, would eventually manage to fully impose monotheism.

Instead, it was a people who lived on the margins of empires—or, rather, squashed between the two great empires—who seized this tension toward monotheism as an opportunity. The genius of Moses, for Gauchet, lay in having captured this tension and dared to overturn the normal power relationships among gods, which were directly dependent on the power relationships among their respective peoples. Israelite tribes were probably in Egypt during the failed attempt by Amenophis to impose monotheism. Less than a century later, Moses picked up this idea of a "supergod," but—this is the novelty— applied it to a god independent of imperial power. He made this god a powerful weapon in his people's resistance by separating his assumed power from the political weakness of its people. Thus armed, Israel won its freedom from exile in Egypt and, later, Babylonia. This "supergod" was not just the god of Israelites—it was a distant god, distant like the emperor, and like the emperor he reigned over all of his peoples; but, like the emperor, he did not love all people equally.

The Israelites became the custodians of the idea of monotheism, and they nurtured it notwithstanding the implicit contradiction between the notion of a universal god and the notion of a chosen people. The contradiction was tentatively resolved in the messianic expectation of a great leader who would finally bring about Israel's dominion over all nations, reestablishing the identity between the god's superiority and the power of the god's people. But history unfolded differently. The long process of unifying the Mediterranean world under a single power was finally achieved, but under Rome, not Israel.

The vast and eventually stable Roman Empire further weakened ancient paganism. What remained was the solitude of individual subjects within the immense empire. The small groups that had formed the structure of humanity, each settled around its own local god, had lost their role as depositaries of identity, legitimacy, power, and knowledge. Within the great empire, to make a mark in one's own city became irrelevant: one had to go

to Rome. The strong sense of identity, which belonging
to a particular group had ever reassured humanity, was
lost.

It was Jesus, in Gauchet's view, who confronted this
deep novel unease and, at the same time, resolved the
contradiction in the Israelites' belief in a universal god
and a chosen people. Jesus repeated Moses's brilliant
overturn, but even more spectacularly: he separated
even more radically religion from power. Jesus and his
follower Paul of Tarsus taught that there was one "true
god," who belonged to everyone but was utterly separate
from imperial power. Jesus created a parallel universe
("My kingdom is not of this world"), in which the scale
of values was inverted with respect to the political world,
and in which a universal god was at the same time remote
and accessible to each individual, without political medi-
ation. A new sphere was born: the sphere of individual
spirituality. The sphere that would be so marvelously
expanded and explored by Saint Augustine. The new
kind of identity was neither social nor political, but
rather personal and individual. The church came into
being as an entity parallel to the world of politics and
assumed its role in helping human beings make sense of
the world. A new, parallel space of identity for each individual was created, separated from the social identity.

But the political power hurried to fill the separation,
and rapidly absorbed the new source of legitimacy: the
Roman Empire became Christian. Godless power had
no choice but to join forces with a god without political
power, and the theocratic foundations of society were
reestablished, at last monotheistic. Still, the rift had been
opened, and it endures. The core of individual spirituality that would give rise to the modern world had been
established.

More recent studies on the origins and nature of religion have underscored the close interdependence of religion and language. They tend to place the origins of religion farther back in time and accentuate the central role religion may have played in the birth of the human race.

In *Ritual and Religion in the Making of Humanity*, a work of vast scope, Rappaport identifies ritual activity as not only the heart of religiosity shared by all cultures, but the very activity around which civilization and humanity* itself grew. It is a profound and well-articulated thesis, supported by a vast body of anthropological evidence, that deserves attention.

Rappaport sees in the ritual function the central axis for the grounding and unraveling of the system of legitimacy at the basis of social life, and even for the reliability of human language itself. Human society coalesces around rites. Ritual activities exist in the animal kingdom, where they generally serve the purpose of communication. Among humans, according to Rappaport, language grows out of these activities.

During the enactment of rites, basic formulas are repeated again and again. Rappaport calls these "Ultimate Sacred Postulates":

Christianity: *Credo in unum Deum*

"I believe in one God."

Islam: أن لا إله إلا الله و أشهد أن محمد رسول الله

"Allah is great, and Mohammed is his Prophet."

---

*In the three possible meanings of "humanity": a particular race of animals; the entirety of characteristics that distinguish human animals from all others; and an ethical value.

Judaism: שְׁמַע יִשְׂרָאֵל יְהֹוָה אֱלֹהֵינוּ יְהֹוָה אֶחָד

"Hear, O Israel: the Lord is our God, the Lord is One."

Or the formula that appears in every prayer of the elaborate Navajo ceremonial:

Growing, we shall walk in beauty and harmony.

Or the great sacred syllable of Hinduism, Jainism, Buddhism, and the Sikh religion, the syllable that contains everything:

ॐ

Om

(Some of these translations are not perfect, but they are the best-known versions of these formulas.) These statements, or the beliefs they refer to, can be neither verified nor falsified. Strictly speaking, they mean nothing. But as they are repeated again and again during rites, they acquire a valence of certainty and rise to become cornerstones upon which to anchor the sacredness from which all of the thought that gives order to the world and legitimacy to the social sphere unfolds.

The key point for understanding what this means is to observe that language does not just reflect reality. Most often, it creates reality. The priest who says, "I now pronounce you man and wife," the judge who says "guilty," the faculty that says, "I now bestow upon you the title of doctor," a parliament that approves a law, Napoléon speaking of honor and glory to French soldiers in the shadow of the pyramids, an American presidential candidate delivering his vision for the future, a minister preaching in church on Sunday—these people do not just describe reality; they make something real by means of language. The higher functions of social life take place in this space created by language. Marriage, citizenship, honesty; doctor, professor, colonel; to be the capital of the United States: these are all realities only inasmuch as they are determined by linguistic enunciations pronounced by members of a society authorized (by whom?) to make them so. Everything having to do with law, honor, institutions, and so on takes place in a space that is largely created and brought into being by language. This space exists only inasmuch as human beings as a group recognize its reality and legitimacy.

The act from which this legitimacy emanates is rite, and the foundation of this legitimacy relies upon the Ultimate Sacred Postulates. These establish a sacred space that consecrates and thus confers legitimacy upon everything that derives from it. The simple act of taking part in a rite implies the recognition of the legitimacy of the meanings that emanate from the rite, quite independently from the intellectual acceptance of specific beliefs enunciated during the rite.

I don't enter this house uninvited, because it's yours. It's yours because you inherited it from your husband. He was your husband because a priest declared him so. The priest was a priest because a bishop ordained him. The bishop was a bishop because the pope elevated him. The pope became pope because God chose him. God exists because, "I believe in one God." And "I believe in one God" because I spoke those words at Mass. Ultimately, I don't enter this house uninvited, because of a binding pact with my peers that is reiterated by my attending Mass. And if I was distracted during Mass and did not believe a word that the priest said, that would not change in the least the global structure to which I nonetheless adhere.

Substituting a judge for a priest, a parliament for the pope, a voting booth or attendance at school for Mass does not change this structure substantially. By enacting their rites again and again, human beings renew the social pact and, at the same time, anchor on a simple gesture their labile and vagabond thoughts about the world.*

THE DIFFERENT FUNCTIONS OF THE DIVINE

This very incomplete sketch of a few theories on the nature of religion offers only an idea of the complexity of this problem—and of our ignorance about these matters. The truth probably lies in some combination of these theories, or in a more complicated narrative, hard to reconstruct.

It seems clear that, one way or another, religious thought was intimately bound up with the working of our logical and mental universe and, most of all, with the social context.

We should not forget that while humans may have been speaking for more than one hundred thousand years, only in the past six thousand years have they left written traces of what they say. We may never know what human beings were saying to one another in the previous hundred millennia, what conceptual structures they tried out, and how many times they changed their minds and started over from scratch. Or maybe someday we will learn something more about this story, and if we do, we will probably find surprises.

---

*Most of the oldest texts we have are about rites. In one of the oldest texts of Indian philosophy, the *Brhadaranyaka Upanishad*,[7] the initial verses—"The head of the sacrificial horse is the dawn. . . . / The sacrificial horse is the world"—make the connection between the rite (the horse that is sacrificed in ancient Hindu ritual) and the world explicit.

The essential point, I think, is that we do not know how and why we think what we think. We do not understand the complexity of the processes that give rise to our thoughts and emotions. Our body, which generates and expresses thoughts and emotions, is a vastly complicated organism, and our ability to understand it is limited. Its complexity is amplified by the fact that none of us exists in solitude. Perhaps our thoughts need to be seen as reflections, upon an individual, of processes that happen at a societal level. We do not think: thoughts pass through us. Asking how we manage to think a given thought may be analogous to asking how a stone in a river manages to raise a wave above it on the water surface.

"Consciousness," "free will," "spirituality," "divinity": these words may be nothing more than ways of indicating processes within ourselves whose complexity escapes us. I believe that this simple realization, which goes back to Baruch de Spinoza, is the most trustworthy compass for making our way through the dark forest of our own thoughts.

We have learned to unveil many of our own mistaken ideas. Twenty-six centuries after Anaximander's first inklings, we have learned to mistrust those who claim to know with certainty that Zeus sends thunderbolts, and analogous contemporary ideas. But we do not yet know how our own mind works. When we seek a sure foundation on which to base decisions about our actions and thoughts, we find that a sure foundation does not exist. We do not even know whether we actually need such a foundation. We continue to make use of vague, uncertain ideas, precisely in those areas that most deeply concern us. What we call "irrational" is the code name for what we don't understand well about ourselves given the limits of our own intelligence.

But this does not imply that we cannot or must not trust our own thinking. To the contrary: our own thinking is the best tool we have for finding our way in this world. Recognizing its limitations does not imply that it is not something to rely upon. If instead we trust in "tradition" more than in our own thinking, for instance, we are only relying on something even more primitive and uncertain than our own thinking. "Tradition" is nothing else than the codified thinking of human beings who lived at times when ignorance was even greater than ours.

The evidence we have from the past few millennia shows slow changes in the form of human thinking, some of which are still unfolding. Ancient polytheism was fairly similar around the Mediterranean and in China, India, Mexico, and South America. Its close relation to social groups and near-identity with political power are similar in all these places as well. A grand path, a great movement, can be glimpsed in the evolution from this polytheism toward the naturalistic-rational openings and democratic institutions of antiquity and their modern reincarnations, passing through the restoration of theocratic monotheism in the late Roman Empire, the Middle Ages, and the Islamic world.

A historical process of immense breadth is unfolding, during which the role of religion in human society has been evolving. This evolution is better measured in millennia rather than centuries. It parallels profound changes not only in the social, political, and psychological structures of society, but also in the forms in which human beings develop knowledge and think about themselves. Anaximander's naturalistic proposal is one chapter in this much larger story.

Let me return then to where I started: the relationship between Ionic thought about the world and religion, and the distinction between the cognitive and other functions of religion. Thales and Anaximander do not explicitly question religion as such. They just omit all mention of stories about the gods. More importantly, they are willing to disregard all certainties, including those inscribed in Rappaport's Ultimate Sacred Postulates. They understand that uncritical acceptance is the post to which we are bound, the cornerstone of our ignorance that prevents us from setting off to explore elsewhere and seeking something truer.

But Thales, smiling, sacrifices a bull to Zeus. Is it thus possible to separate the different functions of religion? Can religion perform its psychological and social functions without being a fundamental obstacle to knowledge? Can we leave a space to those functions, which for centuries were the domain of religions structured within a scaffolding of beliefs, without having to accept the useless burden of wrong ancient beliefs?

Today's religions are not all alike in this respect. There is a continuous spectrum of different attitudes regarding knowledge and a continuous spectrum of intelligence. The spectrum ranges from evangelical groups who believe that the world is no more than six thousand years old, and Catholic dogma; to the antidogmatic impulse of the Unitarian faith; all the way to Buddhism, which describes its own teachings as illusory. Within every religious tradition there is a hide-and-seek game where religious truths beginning to appear nonsensical are immediately reinterpreted in more abstract terms. The god with a long white beard becomes a faceless personal god, then a spiritual principle, then something ineffable of which nothing can be said.

To be sure, not believing that a god is listening to me does not stop me from turning to the sea every morning with a song in my heart, to thank the world for its beauty. There is no contradiction between rejecting irrationalism and listening to the voices of trees, talking to them, touching them, and feeling the flow of their peaceful strength. Trees don't have souls; neither, I think, does my dearest friend. This does not stop me from confiding my emotions to him, talking to trees, and feeling profound joy from these exchanges, giving of my own heart to try to soothe the pain of a friend who is suffering—or watering a parched tree.

To see the sacredness of life and the world, we have no need of a god. We have no need of overworldly guarantors to know that we have values, and even to be willing to die defending them. And if we discover that our generosity and love for others are caused by things found deep inside the recesses of our species' evolutionary process, this does not mean that we love our children and neighbors any less. If the beauty and mystery of things leave us breathless, breathless we remain, moved and in silence.

We have no guarantee for our knowledge. Do we need one? We have no guarantee even about the simple world we see. One hundred micrograms of lysergic acid diethylamide (LSD) suffice for us to see this world in a profoundly different manner. Perhaps not a more or less true way of seeing the world, just a different one. Given our puny knowledge, we can't not accept living in the midst of mystery. It is precisely because mystery exists and is so great around us that those who claim to hold the keys to the mystery cannot be trusted.

Accepting uncertainty and taking a path that seeks constantly new approaches to knowledge imply new risks. A civilization that abandons tried-and-true ways leaves itself open to new dangers. The risk for humanity due to the overheating of our planet that has followed the Industrial Revolution is serious. But traditional ways protect us from these risks even less—they actually make them more uncontrollable. Great civilizations of the past, including the Maya, classical Greece, and perhaps the Roman Empire, have been weakened or altogether wiped away by huge ecological imbalances that they themselves had set in motion, but with the aggravating circumstance that, unlike us, they had no means for understanding what was happening and trying to protect themselves. Intelligence does not necessarily save us from disasters, but it is our best defense.

Henri Bergson wrote that religion was society's defense against the caustic power of intelligence.[8] But who would save us from the caustic power of ignorance? Was the Mayan world saved by its faith in Q'uq'umatz, the serpent god who created humanity? Were the Aztecs saved by Huitzilopochtli, the sun god? Gregory Bateson claims that rational knowledge is necessarily selective, partial, and incapable of understanding the whole.[9] This is certainly true, but this is the mark of every human undertaking—even more so if irrational. It is only by recognizing our limitation and integrating all our tools that we can seek the best paths.

A common misapprehension behind the current powerful antirational tendency is the assumption that rational behavior is selfish, and that only by taming rationality can we come to identify with shared goals and behave in a social, generous manner. I think there is a serious mistake in this. Why on earth should selfish behavior be more rational? The drive to satisfy personal needs may be inscribed in our genetic makeup, but equally so are generosity and sociability. Receiving a gift makes us happy, but giving a gift may make us even happier. Being richer may make us happier, but living in a society without poverty can make us happier still. The assumption that humanity's fundamental motivations are selfish and antagonistic toward others is not rational: it is blind to humanity's complexity.

Conversely, where generosity is concerned, irrational impulses are hardly exemplary. Pure irrationality, the spirit of "wholeness" and "community" that many today see as ballasts of civilization, nourished the rise of Nazi ideology in 1930s Germany, not a champion of generosity. Centuries earlier, thousands of European women were burned at the stake out of an honest irrational desire to save souls.

Thirty centuries ago, human beings, thanks to a course of events that is unknown to us, had surrounded themselves with a system of thought based upon sacrosanct truths. To protect these truths, humanity had developed a complex system of rules, taboos, and power relationships. But reality is change, and the flow of centuries has radically transformed humankind's political, mental, and conceptual structures. We no longer need to worship a pharaoh in order to endow with legitimacy the political systems through which we govern ourselves. There are other ways. We no longer need to invoke Zeus to make sense of thunder and rain. Human beings have built the modern world by accepting uncertainty. This world is the fulfillment of the free dreams of women and men before us. The future can be born only of our own free dreams, but to build the future we need to let go of the past.

Anaximander embodies a step in this process of freeing ourselves from ancient forms of thought. We do not know where this step is leading us. His true discovery was not where rainwater comes from; it was that our ideas can be, and often are, mistaken.

The world is infinitely more complicated than the naïve images we create to find our way through it. The same is true of our thought. The very distinction between the world and our thought is an enigma. Our emotional, social, and psychological complexity far exceed our grasp. We must choose between hiding away in empty truths or accepting the radical uncertainty of our knowledge—remaining, like the Earth, suspended in a void. This choice means trusting in a way of knowing that is keen, effective, but without infallible basis. Only in this way can we continue to understand, recognize our errors and naïveté, broaden our knowledge, and give life the freedom to flourish and grow. I prefer the path of uncertainty. It seems to me that it teaches us more about the world, it is more worthy, more honest, more serious, and more beautiful.

One of Indian civilization's most ancient and fascinating texts, the *Rigveda*, written around 1500 BCE, teaches:

> Who verily knows . . . whence it was born and whence comes this creation?
>
> The Gods are later than this world's production. Who knows then whence it first came into being?
>
> He, the first origin of this creation, whether he formed it all or did not form it,
>
> Whose eye controls this world in highest heaven,
>
> he verily knows it,
>
> or perhaps he knows not.

CONCLUSION:
Anaximander's Heritage

I have considered the legacy of Anaximander from the point of view of a contemporary scientist, and I have used this analysis to reflect on the nature of scientific thinking.

Anaximander emerges as a giant of human thinking, standing at one of the deep roots of modernity. It was he who created what the Greeks called Περὶ φύσεως ἱστορία *peri phuseos istoria* (hence "physics"), the "inquiry into nature," giving birth to a tradition that would form the basis for the entire scientific revolution to come. Even the literary form of this tradition, a treatise in prose, starts with him. His is the first rational view of the natural world. For the first time, the world of things and their relations is seen as directly accessible by the investigation of thought.

In the words of Daniel Graham, "Anaximander's project, in any case, proved in the hands of his successors a program capable of endless development and, in light of its modern incarnation, productive of the greatest advances in knowledge the world has ever known. In a sense his private project has become the grand quest for knowledge of the world."[1]

Anaximander is the first geographer. The first biologist, contemplating the possibility that living beings evolved over time. He is the first astronomer, making a rational study of the movements of heavenly bodies and seeking to reproduce them with a geometrical model. He is the first to propose two conceptual tools that would prove fundamental to scientific activity: the idea of natural law, guiding the unfolding of events over time and by necessity; and the use of theoretical terms to postulate new entities, hypostases used to make sense of the observable world. More important, Anaximander founds the critical tradition that forms the basis of today's scientific thinking: he follows his master's path while at the same time searching for his master's mistakes.

Finally, Anaximander realizes the first great conceptual revolution in the history of science. For the first time, the map of the world is redrawn in depth. The universality of falling bodies is questioned, and a new image of the world is proposed, where space is not structured in "up" and "down," and the Earth floats free in space. It is the discovery of the worldview that will characterize the West for centuries, the birth of cosmology, and the first great scientific revolution. Even more, it is the discovery that scientific revolutions are possible: in order for us to understand the world, we must be aware that our worldview may be mistaken and we can redraw it.

This is the main characteristic of scientific thinking: what seems most obvious to us about the world can be false. Scientific thinking is, therefore, a continuous quest for novel ways of conceptualizing the world. Knowledge is born from a respectful but radical act of rebellion against what we currently think. This is the richest heritage the West has bequeathed to today's global culture, its finest contribution.

This act of rebellion is a challenge launched by Thales and Anaximander: freeing humanity's understanding of the world from the mythical-religious matrix that had structured thought for thousands of years; considering the possibility that the world is understandable without recourse to one or many gods. This is a new prospect for humanity—one that, twenty-six centuries later, still frightens the majority of women and men on this little planet floating in space.

The path opened by Anaximander, the continuous reenvisioning of the world, is an immense adventure. The frightening aspect of this adventure is recognizing our ignorance. I think that accepting our uncertainty is not only the high road to knowledge—it is also the honest and beautiful choice. Our knowledge, like the Earth, floats in nothingness. Its provisional nature and the underlying void do not make life meaningless; they make it more precious.

We do not know where this adventure is leading, but scientific thinking—continuous critical revision of accepted knowledge, openness to the possibility of rebellion against any belief, the ability to explore new images of the world and create novel ones—represents a major chapter in the slow evolution of the history of humankind. It is a chapter opened by Anaximander; we are immersed in it, curious to see where it may lead.

Figure 19. Earth indeed floats in space, suspended in a void.

Notes

Introduction

1. Carlo Rovelli, "Unfinished Revolution," in *Approaches to Quantum Gravity: Toward a New Understanding of Space, Time and Matter,* Daniele Oriti, ed. (Cambridge: Cambridge University Press, 2009). For a collection of points of view on this problem, see the different contributions in this volume.

2. For a nontechnical introduction to the problem of quantum gravity, see Carlo Rovelli, "Quantum Gravity," in *Handbook of the Philosophy of Science,* vol. 2, *Philosophy of Physics,* John Earman and Jeremy Butterfield, eds. (Amsterdam: Elsevier, 2006). For a more technical discussion, see the first chapter of Carlo Rovelli, *Quantum Gravity* (Cambridge: Cambridge University Press, 2004).

Chapter One: The Sixth Century

1. 2 Kings 23:20, 23:14, 23:16 (King James Bible).

2. Quotations from Hesiod are from Hugh G. Evelyn-White's 1914 translations of *Works and Days* and *Theogony,* accessed April 30, 2011, at the Internet Sacred Text Archive, http://www.sacred-texts.com/.

3. Gen. 1:14 (King James Bible).

4. This and the following tablet translation can be found at *Mesopotamia,* the British Museum, accessed April 29, 2011, http://www.mesopotamia.co.uk/astronomer/explore/enuma.html.

5. James Legge, *The Chinese Classics,* vol. 3 (Hong Kong: Hong Kong University Press, 1960), 15, accessed April 29, 2011, http://www.chinapage.com/confucius/shujing-e.html.

6. Quotations from the *Enuma Elish* are from the 1902 translation by L. W. King at the Internet Sacred Text Archive, http://www.sacred-texts.com.

7. See, for example, Jean Bottéro, Clarisse Herrenschmidt, and Jean Pierre Vernant, *Ancestor of the West: Writing, Reasoning, and Religion in Mesopotamia, Elam, and Greece* (Chicago: University of Chicago Press, 2000), 51.

8. Sappho, fragment 31, author's literal translation of text at http://en.wikipedia.org/wiki/Sappho_31.

9. James T. Shotwell, *An Introduction to the History of History* (New York: Columbia University Press, 1922).

10. Carl Roebuck, "The Early Ionian League," *Classical Philology* 50, no. 1 (Jan. 1955), 26-40.

11. Herodotus, *The Histories* 5, 28, accessed January 3, 2011, http://www.greektexts.com/library/Herodotus/index.html.

12. Robert Hahn, "Proportions and Numbers in Anaximander and Early Greek Thought," in Dirk L. Couprie, Robert Hahn, and Gerard Naddaf, *Anaximander in Context: New Studies in the Origins of Greek Philosophy* (Albany: State University of New York Press, 2003), 73.

13. Geoffrey E. R. Lloyd, *Early Greek Science: Thales to Aristotle* (New York: Norton, 1970); Couprie, et al., *Anaximander in Context*.

14. Claudius Aelianus, *Various History*, ch. 17, translated by Thomas Stanley, accessed April 30, 2011, http://penelope.uchicago.edu/aelian/.

15. See, for example, Marcel Conche, *Anaximandre, fragments et témoignages* (Paris: Presses universitaires de France, 1991), 30-35.

CHAPTER TWO: ANAXIMANDER'S CONTRIBUTIONS

1. Horst Blanck, "Anaximander in Taormina," Mitteilungen des deutschen archäologischen Instituts (römische Abteilung) 104: 507-511.

2. Charles H. Kahn, *Anaximander and the Origins of Greek Cosmology* (New York: Columbia University Press, 1964); Marcel Conche, *Anaximandre, fragments et témoignages* (Paris: Presses universitaires de France, 1991); Couprie, et al., *Anaximander in Context*; Daniel W. Graham, *Explaining the Cosmos: The Ionian Tradition of Scientific Philosophy* (Princeton, NJ: Princeton University Press, 2006).

3. The attribution remains controversial. See Charles H. Kahn, "On Early Greek Astronomy," *Journal of Hellenic Studies* 90 (1970): 101-109.

4. Gerard Naddaf, "Antropogony and Politologony in Anaximander of Miletus," in Couprie, et al., *Anaximander in Context*, 10.

5. Lucio Russo, *The Forgotten Revolution: How Science Was Born in 300 BC and Why It Had to Be Reborn* (Berlin: Springer, 2004).

6. Julian B. Barbour, *Absolute or Relative Motion?*, vol. 1, *The Discovery of Dynamics* (Cambridge: Cambridge University Press, 1989).

CHAPTER THREE: ATMOSPHERIC PHENOMENA

1. Sources quoted here and elsewhere are from Conche, *Anaximandre*, ch. 9.

2. Kahn, *Anaximander and the Origins of Greek Cosmology*, 108.

3. Translation of Aristophanes's *The Clouds* taken, with small changes, from http://classics.mit.edu/Aristophanes/clouds.html.

4. Graham, *Explaining the Cosmos*, 17.

5. Shotwell, *An Introduction to the History of History*, 172.

CHAPTER FOUR: EARTH FLOATS IN SPACE, SUSPENDED IN THE VOID

1. Saint Thomas Aquinas, *Summa Theologica*, part 1, article 1, "Reply to Objection 2." "Sciences are differentiated according to the various means through which knowledge is obtained. For the astronomer and the physicist both may prove the same conclusion: that the Earth, for instance, is round." Accessed April 29, 2011 at New Advent, http://www.newadvent.org/summa/1001.htm.

2. "My conviction is, that the earth is a round body in the centre of the heavens, and therefore has no need of air or any similar force to be a support." Plato, *Phaedo*, 74, accessed April 30, 2011, at Project Gutenberg, http://www.gutenberg.org/catalog/world/readfile?pageno=74&fk_files=1446475.

3. Hippolytus, *Refutatio omnium haereseium*, 1, 6, 2-7. Translated from Greek by Marcel Conche, in *Anaximandre*, 191.

4. This is not the case, however, with Dirk Couprie. See, for instance, his fine article on Anaximander in the *Internet Encyclopedia of Philosophy*, accessed January 3, 2011, www.iep.utm.edu/anaximan/.

5. Aristotle, *On the Heavens*, 2, 13, G2v, accessed January 3, 2011, at the University of Virginia Electronic Text Center, http://etext.virginia.edu/toc/modeng/public/AriHeav.html.

6. Hippolytus, cited in Kahn, *Anaximander and the Origins of Greek Cosmology*, 76.

7. More in detail, the reconstruction of Anaximander's logic can be done as follows from the few elements we have. In our experience, heavy objects fall. The Earth is a heavy object; why doesn't it fall?

Anaximander replies: Because for the Earth, all directions are equivalent. Therefore, all directions are not equivalent for the objects that we see falling. For them, a particular direction exists. What is the particular direction that exists for the objects we see falling, but not for the Earth? It cannot be an absolute "down" as we see in figure 13, because such a universal "down" would hold also for the Earth as well, and the argument would make no sense. The particular direction can only be "toward the Earth." Thus objects on the surface of the Earth fall toward the Earth, not toward a universal "down." The Earth itself, therefore, does not need to fall. This is great science. If in addition we accept the credibility of Hippolytus's testimony about Anaximander, then the argument becomes even more transparent. Nothing dominates the Earth. This implies that all objects that fall are dominated by something. By what? Clearly the Earth.

8. In a recent book appearing after the French publication of this book, Dirk Couprie criticizes the idea that Anaximander could have understood the relativity of the notions of "up" and "down," on the grounds that doing this would require a full alternative theory of falling, and this would only come with Aristotle. Couprie, *Heaven and Earth in Ancient Greek Cosmology* (New York: Springer, 2011), 111-114. The key point here, however, is not whether Anaximander had a complete theory of falling. Like Copernicus, he might not have had one. The key point is that Anaximander understood that the universality of "down-falling" in the sense of the left panel of figure 13 does not apply to the Earth. This is incontrovertible, and this is the momentous step made by Anaximander that has the widest importance for the future development of science. As is clear to scientists active in research, what is hard in science is not to add: it is to give up.

9. Hippolytus, cited in Kahn, *Anaximander and the Origins of Greek Cosmology*, 84-85.

10. Couprie, *Heaven and Earth*, 114.

11. Dirk Couprie, "The Discovery of Space: Anaximander's Astronomy," in Couprie, et al., *Anaximander in Context*, 167.

12. Kahn, *Anaximander and the Origins of Greek Cosmology*, 77.

13. Karl Popper, *Conjectures and Refutations: The Growth of Scientific Knowledge* (New York: Routledge, 1998), 186.

CHAPTER FIVE: INVISIBLE ENTITIES AND NATURAL LAWS

1. E. A. Speiser, "Genesis," *The Anchor Bible*, vol. 1. (Garden City, NY: Doubleday, 1964), 3.

2. Gary Witherspoon, *Language and Art in the Navajo Universe* (Ann Arbor: University of Michigan Press, 1977), 46.

3. See, for instance, Aristotle, *Physics*, book 3, pt. 5, accessed April 29, 2011, at the Internet Classics Archive, http://classics.mit.edu/Aristotle/physics.mb.txt.

4. Simplicius, *Aristotelis Physicorum Libros Commentaria*, 24, cited in Conche, *Anaximandre*, 136.

5. Simplicius, *Aristotelis*, cited in Kahn, *Anaximander and the Origins of Greek Cosmology*, 166 (slightly edited).

6. Michael Faraday, *Experimental Researches in Electricity* (London: Bernard Quaritch, 1855), 436-437.

7. Marc Cohen considers an analogous interpretation of the *apeiron* as the first "theoretical entity." Marc Cohen, "History of Ancient Philosophy," lecture notes for University of Washington course Philosophy 320: Ancient Philosophy, accessed January 3, 2011, http://faculty.washington.edu/smcohen/320/320Lecture.html.

8. Iamblichus Chalcidensis, *Life of Pythagoras*, trans. by T. Taylor (London: Inner Traditions/Bear, 1987), vol. 1, sec. 2.

9. The oldest documentation of these words is in a note by an anonymous scholiast of the sixth century CE, possibly identified as the orator Sopatros, in the margin of a manuscript of Aelius Aristides. The story is repeated and used by the sixth-century CE Alexandrian Neoplatonic philosophers Philoponus, Olympiodorus, Elias, and David. The most commonly cited source is the Byzantine John Tzetzes, from the twelfth century CE. David Fowler, *The Mathematics of Plato's Academy: A New Reconstruction* (Oxford: Clarendon Press, 1999), ch. 6.1.

10. According to Simplicius: "Plato assigned circular, regular and ordered motions to the heavens, and offered this problem to the mathematicians: 'which hypotheses of regular, circular and ordered motion are capable of saving the phenomena of the planets?', and first Eudoxus of Knidos produced the hypothesis of the so-called unrolling spheres." Simplicius in *De Caelo*, 492.31, cited in Andrew Gregory, "Eudoxus, Callippus and the Astronomy of the *Timaeus*," *Bulletin of the Institute of Classical Studies*, supp. 78 (2003): 20. For a full discussion of the disputed reliability of this account by Simplicius, see the full text of this nice article by Andrew Gregory.

CHAPTER SIX: REBELLION BECOMES VIRTUE

1. Marcus Tullius Cicero, *Academicorum priorum* 2, accessed January 3, 2011, http://individual.utoronto.ca/pking/resources/cicero/acadprio.txt.
2. See G. E. R. Lloyd, *The Ambition of Curiosity* (Cambridge: Cambridge University Press, 2002).

CHAPTER SEVEN: WRITING, DEMOCRACY, AND CULTURAL CROSSBREEDING

1. Jean-Pierre Vernant, *Les Origines de la pensée grecque* (Paris: Presses Universitaires de France, 1962), ch. 2.
2. *Mesopotamia*, the British Museum, accessed April 29, 2011, http://www.mesopotamia.co.uk/astronomer/explore/secret1t.html.
3. Paul Mazon, *Introduction à l'Iliade* (Paris: Les Belles Lettres, 1967), 294.
4. This has been underscored in classic studies, including two splendid ones by Vernant: *Les Origines de la pensée grecque*, and *Mythe et pensée chez les Grecs* (Paris: Librairie François Maspero, 1965).
5. Herodotus, *Histories* 1, 30.
6. Herodotus, *Histories* 2, 143-144.
7. Shotwell, *Introduction to the History of History*, 144-161.

CHAPTER EIGHT: WHAT IS SCIENCE?

1. Francesca Vidotto, "Nuovi linguaggi per una nuova scienza. L'esperienza del teatro a Padova," *Donne, scienza e potere: Oseremo disturbare l'universo?* (Lecce: Comitato Pari Opportunità, 2006), 81-87, accessed at http://siba-ese.unisalento.it/index.php/pariopp/article/view/8065/7308.
2. This is the central thesis of Lucio Russo's splendid book *The Forgotten Revolution*, a work that may exaggerate on certain points but whose overall argument seems convincing to me.
3. Arthur I. Miller, "The Myth of Gauss's Experiment on the Euclidian Nature of Physical Space," *Isis* 63:3 (1972): 345–348.
4. Barbour, *Absolute or Relative Motion?*, 258.
5. Albert Einstein, *Relativity: The Special and General Theory* (New York: Bartleby, 2000), ch. 28, accessed April 30, 2011, http://www.bartleby.com/173/28.html.
6. For a critique of this reading of science, see Paul Feyerabend, *Against Method* (London and New York: Verso, 1993).
7. See, for instance, Telmo Pievani, *In difesa di Darwin: Creazione senza Dio* (Torino, Italy: Einaudi, 2006).

8. John Stuart Mill, "On Liberty," in *Utilitarianism and On Liberty*, Mary Warnon, ed. (Malden, MA: Wiley-Blackwell, 2003), 103.

CHAPTER NINE: BETWEEN CULTURAL RELATIVISM AND ABSOLUTE THOUGHT

1. Lisa Raphals, "A 'Chinese Eratosthenes' Reconsidered: Chinese and Greek Calculations and Categories," *East Asian Science, Technology and Medicine* 19 (2002): 10–61.

2. A detailed comparison of Greek and Chinese science is in G. E. R. Lloyd, *The Ambition of Curiosity* (Cambridge: Cambridge University Press, 2002).

3. Maurice Godelier, *Antropologia, Storia, Marxismo* (Parma, Italy: Guanda, 1970), cited in Anthony Giddens, *A Contemporary Critique of Historical Materialism* (Stanford: Stanford University Press, 1981), 87.

CHAPTER TEN: CAN WE UNDERSTAND THE WORLD WITHOUT GODS?

1. Saint Augustine, *The City of God*, 7.2, accessed January 3, 2011, http://newadvent.org/fathers/1201.htm.

2. Nicola Abbagnano, *Storia della filosofia*, vol. 1 (Torino, Italy: UTET, 2006), 7.

3. Emmanuele Testa, "Legislazione contro il paganesimo e cristianiz-zazione dei templi nei secoli IV e V," *Studium Biblicum Franciscanum* 41 (1991): 311-326, accessed January 3, 2011, http://198.62.75.1/www1/ofm/sbf/SBFla91.html#Target10. This text on the violent imposition of Christianity over the pagan world is particularly reliable, coming from the Christian cultural world itself.

CHAPTER ELEVEN: PRESCIENTIFIC THOUGHT

1. Joseph Campbell, *Renewal Myths and Rites of the Primitive Hunters and Planters* (Ascona, Switzerland: The Eranos Foundation and Spring Publications, 1989), 2.

2. Julian Jaynes, *The Origin of Consciousness in the Breakdown of the Bicameral Mind* (Boston: Houghton Mifflin, 1976).

3. Roy A. Rappaport, *Ritual and Religion in the Making of Humanity* (Cambridge: Cambridge University Press, 1999).

4. Emile Durkheim, *The Elementary Forms of the Religious Life* (Oxford: Oxford University Press, 2001), 154.

5. Jaynes, *Origin of Consciousness*, ch. 4.

6. Marcel Gauchet, *Le Désenchantement du monde* (Paris: Gallimard, 1985).

7. *The Thirteen Principal Upanishads*, trans. Robert Ernest Hume (Oxford: Oxford University Press, 1931).

8. Henri Bergson, *Les deux sources de la morale et de la religion* (1935), accessed January 3, 2011, http://classiques.uqac.ca/classiques/bergson_henri/deux_sources_morale/deux_sources_morale.html.

9. Gregory Bateson, *Steps to an Ecology of Mind: Collected Essays in Anthropology, Psychiatry, Evolution, and Epistemology* (New York: Ballantine, 1972).

Conclusion: Anaximander's Heritage

1. Graham, *Explaining the Cosmos*, 17.

BIBLIOGRAPHY

GENERAL RESOURCES

A rather complete bibliography on Anaximander is in *Couprie, Dirk. "Bibliography on Anaximander,"* accessed January 3, 2011, http://www.dirkcouprie.nl/Anaximander-bibliography.htm.
For an overview of the history of Miletus and a rich bibliography, see http://www.ehw.gr/asiaminor/forms/fLemmaBody Extended.aspx?lemmaID=8177.

PUBLISHED RESOURCES

Abbagnano, Nicola. *Storia della filosofia.* Torino, Italy: UTET, 2006.
Aristophanes. *The Clouds.* Accessed January 3, 2011, http://classics.mit.edu/Aristophanes/clouds.html.
Aristotle. *On the Heavens.* Accessed April 29, 2011, http://etext.virginia.edu/toc/modeng/public/AriHeav.html.
————. *Physics.* Accessed April 29, 2011, http://classics.mit.edu/Aristotle/physics.mb.txt.
Augustine of Hippo, Saint. *The City of God.* Accessed January 3, 2011, http://newadvent.org/fathers/1201.htm.
Barbour, Julian B. *Absolute or Relative Motion?* Vol. 1, *The Discovery of Dynamics.* Cambridge: Cambridge University Press, 1989. This is a splendid overview of the history of science, and of astronomy and physics in particular, up to Newton.

Barnes, Jonathan. *The Presocratic Philosophers*. London: Routledge & Kegan Paul, 1979.

Bateson, Gregory. *Steps to an Ecology of Mind: Collected Essays in Anthropology, Psychiatry, Evolution, and Epistemology*. New York: Ballantine, 1972.

Bergson, Henri. *Les deux sources de la morale et de la religion* (1935). Accessed January 3, 2011, http://classiques.uqac.ca/classiques/bergson_henri/deux_sourc es_morale/deux_sources_morale.html.

Blanck, Horst. "Anaximander in Taormina." Mitteilungen des deutschen archäologischen Instituts (römische Abteilung) 104 (1997): 507-511.

Bottero, Jean, Clarisse Herrenschmidt, and Jean Pierre Vernant. *Ancestor of the West: Writing, Reasoning, and Religion in Mesopotamia, Elam, and Greece*. Chicago: University of Chicago Press, 2000. Three penetrating essays on the culture of the ancient Middle East.

Campbell, Joseph. *Renewal Myths and Rites of the Primitive Hunters and Planters*. Ascona, Switzerland: The Eranos Foundation and Spring Publications, 1989.

Cicero, Marcus Tullius. *Academicorum priorum 2*. Accessed January 3, 2011, http://individual.utoronto.ca/pking/ resources/cicero/acadprio.txt.

Claudius Aelianus. *Various History*. Translated by Thomas Stanley. Accessed April 30, 2011, http://penelope.uchicago.edu/aelian/.

Cohen, Marc. "History of Ancient Philosophy," lecture notes for University of Washington course Philosophy 320: Ancient Philosophy, accessed January 3, 2011, http://faculty.washing- ton.edu/smcohen/320/320Lecture.html.

Colli, Giorgio. *La Sagesse grecque: Epiménide, Phérécyde, Thalès, Anaximandre*. Tome 2. Paris: Editions de l'éclat, 1992.

Conche, Marcel. *Anaximandre, fragments et témoignages*. Paris: Presses universitaires de France, 1991. An ample collection of ancient sources, with a critical discussion.

Couprie, Dirk L. "Anaximander." The Internet Encyclopedia of Philosophy. Accessed January 3, 2011, http://www.iep.utm.edu/anaximan/. A fine introduction, with many ideas similar to those expressed here.

———. *Heaven and Earth in Ancient Greek Cosmology*. New York: Springer, 2011.

———. "The Visualization of Anaximander's Astronomy." *Apeiron* 28 (1995): 159-181.

Couprie, Dirk L., Robert Hahn, and Gerard Naddaf. *Anaximander in Context: New Studies in the Origins of Greek Philosophy*. Albany: State University of New York Press, 2003.

Diels, Hermann Alexander, and Walter Krantz. *Die Fragmente der Vorsocratiker*. Berlin: Weidmannsche, 1951. The main complete collection of primary sources on pre-Socratic philosophers.

Diogenes Laertius. *Vitae philosophorum*. Oxford: Oxford University Press, 1964.

Dumont, Jean-Paul, Daniel Delattre, and Jean-Louis Poirier. *Les Présocratiques*. Paris: Gallimard, 1988.

Durkheim, Emile. *The Elementary Forms of the Religious Life*. Oxford: Oxford University Press, 2001.

Earman, John, and Jeremy Butterfield, eds. *Handbook of the Philosophy of Science*. Vol. 2. *Philosophy of Physics*. Amsterdam: Elsevier, 2006.

Einstein, Albert. *Relativity: The Special and General Theory*. New York: Bartleby, 2000. Accessed April 30, 2011, http://www.bartleby.com/173/.

Eliade, Mircea. *Trattato di storia delle religioni*. Torino, Italy: Bollati Boringhieri, 1999. Another classic in the interpretation of religious phenomena.

Enuma Elish. Translated by L. W. King, 1902. Accessed April 30, 2011, http://www.sacred-texts.com/ane/enuma.htm.

Faraday, Michael. *Experimental Researches in Electricity*. London: Bernard Quaritch, 1855.

Farrington, Benjamin. *Science in Antiquity*. London: T. Butterworth, 1936.

Feyerabend, Paul. *Against Method*. London and New York: Verso, 1993.

Feynman, Richard P. "The Development of the Space-Time
    View of Quantum Electrodynamics." 1965 Nobel Lecture.
    Accessed January 3, 2011, http://nobelprize.org/nobel_prizes/
    physics/laureates/1965/feynman-lecture.html.
Fowler, David. *The Mathematics of Plato's Academy: A New*
    Reconstruction. Oxford: Clarendon Press, 1999.
Gauchet, Marcel. *Le Désenchantement du monde.* Paris:
    Gallimard, 1985.
Giddens, Anthony. *A Contemporary Critique of Historical*
    Materialism. Stanford: Stanford University Press, 1981.
Godelier, Maurice. *Antropologia, Storia, Marxismo.* Parma, Italy:
    Guanda, 1970.
Graham, Daniel W. *Explaining the Cosmos: The Ionian Tradition*
    of Scientific Philosophy. Princeton, NJ: Princeton University
    Press, 2006. A very good recent presentation of the tradition
    of scientific philosophy in the Ionian school.
Gregory, Andrew. "Eudoxus, Callippus and the Astronomy of
    the Timaeus." Supplement 78, *Bulletin of the Institute of*
    Classical Studies (2003): 5-28.
Guthrie, W. K. C. *The Earlier Presocratics and the Pythagoreans.*
    Vol. 1 of *A History of Greek Philosophy.* Cambridge: Cambridge
    University Press, 1962.
———. *The Presocratic Tradition from Parmenides to Democritus.*
    Vol. 2 of *A History of Greek Philosophy.* Cambridge: Cambridge
    University Press, 1965.
Herodotus. *Histories.* Accessed January 3, 2011,
    http://www.greektexts.com/library/Herodotus/index.html.
Hesiod. *Theogony.* Translated by Hugh G. Evelyn-White, 1914.
    Accessed April 30, 2011, http://www.sacred-
    texts.com/cla/hesiod/works.htm.
———. *Works and Days.* Translated by Hugh G. Evelyn-White,
    1914. Accessed April 30, 2011,
http://www.sacred-texts.com/cla/hesiod/works.htm.
Iamblichus Chalcidensis. *Life of Pythagoras.* Translated by T.
    Taylor. London: Inner Traditions/Bear, 1987. Also, accessed
    April 29, 2011, http://www.completepythagoras.net/main-
    frameset.html.

Jaynes, Julian. *The Origin of Consciousness in the Breakdown of the Bicameral Mind.* Boston: Houghton Mifflin, 1976. A controversial hypothesis about the role of the divine in the founding of civilization.

Jeannière, Abel. *Les Présocratiques: l'aurore de la pensée grecque.* Paris: Seuil, 1996.

Kahn, Charles H. *Anaximander and the Origins of Greek Cosmology.* New York: Columbia University Press, 1964. Original source texts about Anaximander and a stringent critical evaluation of their reliability.

———. "On Early Greek Astronomy." *Journal of Hellenic Studies* 90 (1970): 101-109. Discusses the attribution to Anaximander of the measurement of the ecliptic.

Kirk G. S., J. E. Raven, and M. Schofield. *The Presocratic Philosophers: A Critical History with a Selection of Texts.* Cambridge: Cambridge University Press, 1983.

Lahaye, Robert. *La Philosophie ionienne, l'Ecole de Milet: Thalès, Anaximandre, Anaximène, Héraclite d'Éphèse.* Paris: Editions du Cèdre, 1966.

Legge, James. *The Chinese Classics.* Vol. 3. Hong Kong: Hong Kong University Press, 1960. Accessed April 29, 2011, http://www.chinapage.com/confucius/shujing-e.html.

Legrand, Gérard. *Les Présocratiques.* Paris: Bordas, 1987.

Lloyd, Geoffrey E. R. *The Ambition of Curiosity.* Cambridge: Cambridge University Press, 2002. An interesting comparison of ancient Greek and Chinese knowledge.

———. *Early Greek Science: Thales to Aristotle.* New York: Norton, 1970. A brief classic on Greek science.

Mazon, Paul. *Introduction à l'Iliade.* Paris: Les Belles Lettres, 1967.

McLennon, J. "Shamanic Healing, Human Evolution, and the Origin of Religion." *Journal for the Scientific Study of Religion* 36 (1980): 345-354. An evolutionary approach to religious anthropology.

Mill, John Stuart. "On Liberty," in *Utilitarianism and On Liberty.* Edited by Mary Warnon. Malden, MA: Wiley-Blackwell, 2003.

Miller, Arthur I. "The Myth of Gauss's Experiment on the Euclidian Nature of Physical Space." *Isis* 63:3 (1972): 345–348.

Oriti, Daniele, ed. *Approaches to Quantum Gravity: Toward a New Understanding of Space, Time and Matter*. Cambridge: Cambridge University Press, 2009.

Pievani, Telmo. *In difesa di Darwin: Creazione senza Dio*. Torino, Italy: Einaudi, 2006.

Plato. *Phaedo*. Accessed April 30, 2011, http://www.gutenberg.org/catalog/world/readfile?fk_files=144 6475.

Popper, Karl. *Conjectures and Refutations: The Growth of Scientific Knowledge*. New York: Routledge, 1998.

———. *The World of Parmenides: Essays on the Presocratic Enlightenment*. Edited by Arne F. Peterson and Jørgen Mejer. New York: Routledge, 1998.

Raphals, Lisa. "A 'Chinese Eratosthenes' Reconsidered: Chinese and Greek Calculations and Categories." *East Asian Science, Technology and Medicine* 19 (2002): 10–61.

Rappaport, Roy A. *Ritual and Religion in the Making of Humanity*. Cambridge: Cambridge University Press, 1999. A great classic on the nature of religion.

Reynolds, Vernon, and Ralph Tanner. *The Social Ecology of Religion*. New York: Oxford University Press, 1995.

Robinson, John Mansley. *An Introduction to Early Greek Philosophy*. Boston: Houghton Mifflin, 1968.

Roebuck, C. "The Early Ionian League." *Classical Philology* 50, no. 1 (Jan. 1955): 26-40.

Rovelli, Carlo. *Quantum Gravity*. Cambridge: Cambridge University Press, 2004.

———. "Quantum Gravity," in *Handbook of the Philosophy of Science*, Vol. 2, *Philosophy of Physics*. Edited by John Earman and Jeremy Butterfield. Amsterdam: Elsevier, 2006.

———. "Unfinished Revolution," in *Approaches to Quantum Gravity: Toward a New Understanding of Space, Time and Matter*. Edited by Daniele Oriti. Cambridge: Cambridge University Press, 2009.

Russo, Lucio. *Flussi e riflussi*. Milano: Feltrinelli, 2003. A short, keen treatise on ancient knowledge about tides and its possible influence on the scientific renaissance in the seventeenth century.

———. *The Forgotten Revolution: How Science Was Born in 300 BC and Why It Had to Be Reborn*. Berlin: Springer, 2004. An important and passionate work by a mathematician that brings together an immense amount of information on Alexandrian science, shedding light on its complexity and richness. The work shows clearly how lack of mathematical and scientific competence easily leads to misunderstanding or underestimation of ancient science. This is an important text for understanding ancient science.

Sappho. Accessed April 30, 2011, http://en.wikipedia.org/wiki/Sappho_31.

Shotwell, James T. *An Introduction to the History of History*. New York: Columbia University Press, 1922.

Smolin, Lee. *The Life of the Cosmos*. New York: Oxford University Press, 1997.

Speiser, E. A. *Genesis*. Translation, introduction, and notes. Vol. 1 of the Anchor Bible Series. Garden City, NY: Doubleday, 1964.

Testa, Emmanuele. "Legislazione contro il paganesimo e cristianizzazione dei templi nei secoli IV e V." *Studium Biblicum Franciscanum* 41 (1991). Accessed January 3, 2011, http://198.62.75.1/www1/ofm/sbf/SBFla91.html#Target10.

The Thirteen Principal Upanishads. Translated by Robert Ernest Hume. Oxford: Oxford University Press, 1931.

Unger, Roberto Mangabeira. *The Self Awakened: Pragmatism Unbound*. Cambridge, MA: Harvard University Press, 2007. A fascinating manifesto for a philosophy and a politics in never-ending evolution.

Vernant, Jean-Pierre. *Mythe et pensée chez les Grecs*. Paris: Librairie François Maspero, 1965.

———. *Les Origines de la pensée grecque*. Paris: Presses Universitaires de France, 1962. A brief classic covering the relationship between the specificity of Greek political organi-

zation and the originality of Greek thought. Includes a fine
reconstruction of the Mycenaean cultural world and the evolution of political structures in the Greek world.

Vidotto, Francesca. "Nuovi linguaggi per una nuova scienza.
L'esperienza del teatro a Padova." *Donne, scienza e potere:
Oseremo disturbare l'universo?* Lecce: Comitato Pari
Opportunità, 2006: 81-87. http://siba-
ese.unisalento.it/index.php/pariopp/article/view/8065/7308.

Witherspoon, Gary. *Language and Art in the Navajo Universe.*
Ann Arbor: University of Michigan Press, 1977.

INDEX

Image Credits

Acknowledgments

I thank Fabio Soso for handing down to me a passion for ancient science; Dirk Couprie, one of the greatest experts on Anaximander, for reviewing these pages with patience and correcting my worst errors; Barbara Goio, for her excellent suggestions and keen editing; Lucinda Bartley, for her extensive and precious editorial work on the English edition; and my parents, for so much more.